AF744187

Analyzing Strategic Behavior in Business and Economics

Analyzing Strategic Behavior in Business and Economics

A Game Theory Primer

Thomas J. Webster

LEXINGTON BOOKS
Lanham • Boulder • New York • Toronto • Plymouth, UK

£70
LIBRARY
University of Nottingham
WITHDRAWN
8142461
658
403
3
WEB

Published by Lexington Books
A wholly owned subsidiary of Rowman & Littlefield
4501 Forbes Boulevard, Suite 200, Lanham, Maryland 20706
www.rowman.com

10 Thornbury Road, Plymouth PL6 7PP, United Kingdom

Copyright © 2014 by Lexington Books

All rights reserved. No part of this book may be reproduced in any form or by any electronic or mechanical means, including information storage and retrieval systems, without written permission from the publisher, except by a reviewer who may quote passages in a review.

British Library Cataloguing in Publication Information Available

Library of Congress Cataloging-in-Publication Data

Webster, Thomas J.
Analyzing strategic behavior in business and economics : a game theory primer / Thomas J. Webster.
pages cm
Includes bibliographical references and index.
ISBN 978-0-7391-8604-6 (cloth) — ISBN 978-0-7391-8605-3 (electronic) 1. Game theory. 2. Decision making. 3. Management. I. Title.
HB144.W33 2014
658.4'033—dc23

2013043736

∞™ The paper used in this publication meets the minimum requirements of American National Standard for Information Sciences—Permanence of Paper for Printed Library Materials, ANSI/NISO Z39.48-1992.

Printed in the United States of America

Contents

Preface

Standard optimization analysis involves finding the most efficient solution to well-defined objective function subject to one or more side constraints. A manager's ability to maximize a firm's value on behalf of shareholders, for example, may be hamstrung by an inability to raise finance capital, shortages of skilled labor, disruptions in the flow of critical raw materials, social or political unrest in overseas markets, capacity constraints, insufficient warehouse space, labor unrest, pollution emission standards, government regulation, proscriptions against certain types of business activities, antitrust legislation, and so on. Unfortunately, this approach to finding optimal solutions to managerial decision-making problems often ignores the competitive realities of the marketplace.

Business decisions are rarely made in a competitive vacuum. Consider, for example, the decision by senior management of a publicly-owned corporation to launch an advertising campaign in the expectation of increasing market share and profits. It commits substantial resources to the expectation of increasing sales, profit, and market share. But, what else should this company factor into its calculations? For one thing, its rivals in the marketplace will not stand idly by and watch their customer switch brand loyalties. They will respond aggressively to recapture, if not expand, their market shares of the market. This countermove may lead to even greater spending by all firms, but no appreciable change in existing market shares. In the end, the bottom lines of all firms may suffer. When the actions of a firm affect, and are affected by, the actions of rivals, it is important to ask whether a managerial decision takes you to a place that you want to be, or that you don't want to be.

Managers who are able to put themselves in the shoes of rivals are more likely to successfully achieve the firm's objectives than those who do not. Managers must be able to distinguish between those situations in which conflict is the best course of action, and those in which the best result entails cooperation. The purpose of this text is to introduce students to the fundamentals of game theory, which is the formal study of strategic behavior. Game theory is not a cookbook of recipes on how to deal with every strategic situation. It is, however, a tool kit for analyzing situations involving move and countermove.

Prerequisites

Most of the economics presented in this book is at the introductory level. For this reason, this book assumes that the reader has taken a course in the principles of microeconomics. More advanced economics courses, such as managerial economics, industrial organization, and international trade are helpful, but by no means mandatory. Chapter 13, for example, explores the possible effects of export subsidies on intraindustry trade. The purpose of this chapter is not to educate the reader in the theories of international trade. Rather, it is meant to show how continuous strategies and best-response functions can enhance our understanding of commercial policy and international trade flows.

Although the presentations of most of the topics presented in this text are diagrammatic, it is not possible to explore some important game theoretic concepts without the use of basic mathematics and statistics. In part II, for example, simple differential calculus is used to identify players' best-response functions. In parts III and IV, such basic statistical concepts as expected value, standard deviation, and conditional probabilities are used in games involving incomplete and imperfect information. On the brighter side, this book is axiomatic in nature and presentation. There are no formal proofs and formal training in real analysis and set theory is not required.

Organization

This book is organized according to the information available to decision makers and the order of play. Sections I and II of the book deal with static and dynamic games in which the players have complete information about the strategies and payoffs. Section III discusses static and dynamic games under conditions of incomplete information in which the payoffs from alternative strategy profiles are uncertain. Section IV examines dynamic games with imperfect information in which the players are uncertain about the prior moves of rivals.

Depending on the target audience and the curriculum, chapters 1–10, 16, and 18 constitute the core of an introductory course in game theory and require little or no formal training in mathematics or statistics. Chapters 11–17 in section I expands the earlier analysis of static games with complete information by introducing the continuous strategies and best response functions. A rudimentary understanding of optimization analysis is required.

Sections III and IV build upon concepts presented earlier in the book. These sections relax the assumptions of complete and perfect information and explore some real-world applications of game theoretic concepts, such as incentive contracts and auctions. The material presented in these chapters requires a familiarity with such basic statistical concepts as risk and uncertainty, which are reviewed in chapter 19. Although somewhat more challenging, these topics can be mastered by most students with a small measure of patience and perseverance.

Note to the Instructor

An Instructors Manual, including answers to selected end-of-chapter exercises and test bank, and PowerPoint presentations is available free of charge to instructors who adopt this text. For more information, contact textbooks@rowman.com.

CHAPTER ONE

~

Introduction to Game Theory

Introduction

Many of us think of a game as a recreational activity involving two or more players, such as chess, gin rummy, football, or the computer game *Call of Duty*, which involves both winners and losers. In business, we often speak of the winning "team" as the company with best return on equity or largest market share. Unlike chess, however, success in business sometimes results from cooperation to achieve a mutually beneficial outcome.

Game theory is the formal study of decision making in strategic situations involving move and countermove. Game theory can deepen our understanding of strategic situations involving win-win situations when cooperation is possible. It also helps us to identify best responses when cooperation is not possible.

How well we perform in situations involving move and countermove depends on several factors, including a player's strategic skills, understanding of the rules, and information about a rival's resources, goals, and intentions. Master chess players, for example, do not simply react to moves, but attempt to unravel an opponent's strategy and tactics. It would be shortsighted for a manager to lower price to boost sales without first anticipating how competitors may react to a loss in market share, revenues, and profits. It would be negligent for a national leader to ignore the potential consequences of a change in foreign policy that invites a terrorist attack or risks plunging a country into war.

Strategic Behavior

Game theory is the most important tool in the economist's analytical kit for analyzing strategic behavior. **Strategic behavior** involves situations in which the decisions of individuals and groups affect, and are affected by, the decisions of other individuals and groups. In other words, the decision-making process is mutually interdependent.

Much of economic theory assumes that decision makers attempt to achieve the best outcome subject to constraints that limit our options. Firms, for example, attempt to maximize shareholder value subject to limited financial, human, and nonhuman resources. Analyzing the decision-making process is simplified when we assume that these decisions have little or no measurable effect on the behavior of rival firms. In the real world, however, this assumption is not just naïve; it can jeopardize a company's very survival.

Game theory represents a significant improvement over earlier attempts to analyze strategic behavior. Although managers have much to learn from game theory, it is not a cookbook that provides a recipe for dealing with every strategic situation. Game theory attempts to systematically analyze situations in which the outcomes are mutually interdependent.

Short History of Game Theory

One of the earliest attempts to analyze strategic behavior can be traced to the work of French economist Antoine Augustin Cournot (28 August 1801–31 March 1877). Cournot (1838) sought to explain the output-setting decisions of duopolies. His analysis was modified five decades later by Joseph Bertrand (11 March 1822–5 April 1900), whose work (Bertrand, 1883) emphasized the role of strategic behavior in product pricing. In the twentieth century, the contributions of Cournot and Bertrand to an understanding of imperfect competition were extended by Edward Chamberlin (18 May 1899–16 July 1967), Joan Robinson (31 October 1903–5 August 1983), Heinrich von Stackelberg (31 October 1905–12 October 1946), Paul Sweezy (10 April 1910–27 February 2004), and others, although their efforts were handicapped by their reliance on standard economic and optimization methodologies.

In the 1920's, French mathematician Émile Borel (7 January 1871–3 February 1956) published several papers on the theory of games. By examining the role of bluffing and second-guessing in poker games with incomplete information, Borel sought to identify the existence of dominant strategies, which he believed would have military and economic applications. Although he is credited with the first modern formulation of mixed strategies and suggested a minimax equilibrium for two-player games with three or five possible strategies, he did not develop his ideas very far. Because of this, most historians recognize the Hungarian-born mathematician John von Neumann (28 December 1903–8 February 1957) as the father of modern game theory.

The brilliant and versatile von Neumann began his work on game theory by modeling the behavior of poker players. In 1928 he proved his famous minimax theorem, which demonstrates that in zero-sum games players with perfect information will adopt a strategy that minimizes their maximum losses. Recognizing the importance of game theory to economics, von Neumann teamed up with German-born economist Oskar Morgenstern (24 January 1902–26 July 1977) to develop his ideas more fully.

The result was the publication in 1944 of the modern game theory classic, *Theory of Games and Economic Behavior*.

Although von Neumann and Morgenstern emphasized the role of cooperation in economic behavior, Princeton mathematician John Forbes Nash Jr. (13 June 1928–) believed that strategic behavior was essentially rivalrous. In his twenty-seven page doctoral dissertation and several papers published in the early 1950s, Nash identified in games with a finite number of strategies the existence a "fixed point" equilibrium in which rational players adopt their best response to the strategies adopted by rivals. This seminal result later became known as a Nash equilibrium. According to Nash, business managers, for example, will cooperate with rivals when it suits them, but will violate implicit or explicit agreements whenever it is in their best interests to do so.

The ideas developed by Nash transformed game theory from a highly focused tool for analyzing zero-sum situations into a more general approach for analyzing strategic behavior involving both competition and cooperation. Although game theory was initially developed for use by economists, its usefulness to such diverse fields as sociology, politics, warfare, and diplomacy became immediately apparent. Game theory has even been applied to problems in evolutionary biology.

Since the 1970s, game theory has fueled a revolution in economic thought. In the 1990s, game theory captured the public's imagination when it was used to design an auction to allocate licenses for the "narrow-band" frequencies used by mobile telecommunication services such as cellular phones, pagers, and other wireless devices. These auctions both efficiently allocated Federal Communications Commission licenses and generated billions of dollars for U.S. government coffers.

Lexicon of Game Theory

In game theory, a decision maker is referred to as a **player**. Players' decisions have consequences, which we refer to as payoffs. A **payoff** is the expected gain or loss that results from a player's strategy choices.

Games can have zero-sum or non-zero-sum payoffs. In a **zero-sum game**, one player's gain is another player's loss, such as wagering on a hand in poker. In a **non-zero-sum game**, the sum of the players' gains and losses does not equal zero. In non-zero-sum games, the players can be all winners or all losers.

There are several features common to all games. Games have rules that define the order of play. Moves are based on strategies. A **strategy** is a game plan for achieving a player's objective. A player's strategy may consist of a single move, or multiple moves. It may involve pure or mixed strategies. A **pure strategy** is a complete and nonrandom game plan in which each player's move is the best response to the actual or anticipated move of a rival. By contrast, a **mixed strategy** involves randomly mixing pure strategies. In mixed strategy games, moves and payoffs are probabilistic.

The complete description of the players' strategies is called a **strategy profile**. In static games, strategy profiles will be summarized within curly brackets. Suppose, for example, that the managers of rival firms A and B are considering charging a *high price* or a *low price*. This game has four possible strategy profiles: {*High price*, *High price*}, {*High price*, *Low price*}, {*Low price*, *High price*}, and {*Low price*, *Low price*}. The first entry refers to the strategy adopted by A and the second entry is the strategy adopted by B. Each strategy profile results in payoffs, which will be summarized in parentheses. Suppose that the payoff to each firm from a *high-price* strategy is $1 million. These payoffs would be summarized as ($1 million, $1 million).

There are two basic types of games. In a **static game**, players do not know the other players' moves until all moves have been made. A type of static game is a **simultaneous-move game** in which all players move at the same time. For example, in the children's game "Rock, paper, scissors" the two players simultaneously show a closed fist (rock), an open hand (paper), or a separated index and middle finger (scissors). The winner is determined by the rule: rock breaks scissors, scissors cut paper, and paper covers rock.

In static games, it is not necessary for players to move at the same time. We could, for example, isolate the players who reveal their strategy choices to an impartial referee, who declares a winner. Only then do the players learn of their opponents' moves.

By contrast, the players in a **dynamic game** (also known as a *sequential-move game*) alternate moves. In a two-player dynamic game, for example, player A moves first, followed by player B, followed by player A or B, and so on. Chess is an example of a sequential-move game.

Finally, games are defined by the number of times they are played. A **one-time game** is played once. A **repeated game** is played multiple times. If you and a friend agree to play a single game of backgammon, you are playing a one-time game. On the other hand, the best two out of three is a repeated game.

Rational versus Actual Behavior

In game theory, we generally assume that players behave rationally. By **rational behavior** we mean that players make decisions that are in their own self-interest. Sometimes the objective is to maximize expected measurable payoffs, such as profits, revenues, and market share, or to minimize expected payoffs, such as jail time following the commission of a crime. Sometimes rational choices are based on maximizing subjective personal satisfaction, such as when an individual receives positive utility from the well-being of others.

In order to make rational choices, players must have complete information about payoffs, strategy choices, and the preferences of other players. Even when decision makers have a shared understanding of the strategic environment within which games are played, preconceptions, misperceptions, and biases influence players' strategy choices. Herbert Simon (1957) recognized the limitations imposed on purely rational behavior in situations characterized by less-than-complete information and computational complexity. Simon noted that decision-makers act both rationally and emotionally depending on their cognitive limitations and the time available to make decisions. Simon characterized this as *bounded rationality*. So, how useful is game theory for understanding real-world behavior?

Game theory seeks to describe a player's best response to situations involving move and countermove. It does not necessarily explain the way decision makers actually behave. As Thomas Schelling (1960) noted:

> A study of conscious, intelligent, sophisticated conflict behavior—of successful behavior—is like the search for rules of "correct" behavior in a contest-winning sense. . . . We may wish to understand how participants actually do conduct themselves in conflict situations; an understanding of the "correct" play may give us a bench mark for the study of actual behavior. (p. 3)

Or, as John Harsanyi (1977) elegantly observed:

> Our theory is a *normative* (prescriptive) theory rather than a *positive* (descriptive) theory. At least formally and explicitly it deals with the question of how each player *should* act in order to promote his own interests most effectively in the game and not with the question of how he (or persons like him) *will* actually act in a game of this particular type. (p. 16)

The payoffs in many of the games presented in this text are represented by pure numbers that reflect the players' preferences. We assume that player preferences are transitive in the sense that if payoff A is preferred to payoff B, and payoff B is preferred to payoff C, then payoff A is preferred to payoff C. Thus, the larger payoffs are associated with more preferred outcomes. In other cases, payoffs are monetary. In this case, we will assume that a dollar yields equivalent utility for all players. Finally, in games where there are multiple equilibria, we will evaluate the best outcome as *Pareto efficient* if it is not possible to make one player better off without making some other player worse off.

Practice Exercises

1.1 What is meant by strategic behavior?

1.2 What is the difference between a player's strategy and a player's move?

1.3 Distinguish between a static game and a simultaneous-move game.

1.4 Provide an example of a zero-sum and a non-zero-sum game.

1.5 Distinguish between a pure strategy and a mixed strategy.

1.6 Explain how the assumption of rational behavior affects a player's strategy choices. What are the limitations of this assumption for predicting actual behavior?

1.7 In the card game "war," a standard deck of cards is shuffled and equally divided between two players. The cards are valued from lowest to highest 2–10, jack, queen, king, and ace. Suits (clubs, diamonds, hearts, and spades) in this game are irrelevant. The players then recite in unison "w-a-r spells war." As each letter is recited, the players place a card face down on the table. On the word "war," a card is placed face up. The player with the highest valued card wins the cards on the table. If two players show the same card, the process is repeated until there is a winner. The game ends when the players have gone through the deck. The player with the greatest number of cards at the end of the game wins. Alternatively, the game could be played by isolating the players in separate rooms. An impartial referee then examines each player's "war" card in turn and declares a winner. Both versions of this game are static, but only one version is simultaneous. Explain.

PART I

STATIC GAMES WITH COMPLETE INFORMATION

CHAPTER TWO

Coalition Games

Introduction

The first part of this book will be devoted to analyses of static games with complete information. In this chapter, we will examine perhaps the best known of all game theoretic scenarios—the *prisoner's dilemma.* The prisoner's dilemma is an example of a **coalition game** in which it is in the players' best interest to cooperate, but where each has an incentive to adopt a strategy that results in a less-than-optimal outcome.[1] By **complete information** we mean that the players, their strategy options, and the payoffs are common knowledge.

Prisoner's Dilemma

The prisoner's dilemma is an example of a two-player, static, coalition game in which each player has a **strictly-dominant strategy**. A strategy is strictly dominant when it is the same best response to any strategy adopted by another player. Consider the following versions of the prisoner's dilemma.[2]

Two suspects are arrested by the police following a burglary. Although the authorities believe them to be guilty, there is not enough evidence to convict them of the crime. If neither suspect confesses, there is only enough evidence to convict them of a misdemeanor, which carries a penalty of six months in jail. A suspect who confesses and turns state's evidence will be set free, while the suspect who remains silent goes to jail for ten years. If both suspects confess and testify against the other, each will receive a five-year jail sentence. To extract a confession, the suspects are separated and advised of their options.

The Extensive Form

The extensive form of the prisoner's dilemma is illustrated in Fig. 2.1. Also referred to as a *game tree*, the **extensive form** summarizes the players, the information available to each player at each stage of the game, and the payoffs from alternative strategy profiles. Although the extensive form is better suited for describing sequential-move games, it has been modified to capture the basic features of a static game. This is important since many game theoretic situations have both static and dynamic elements.

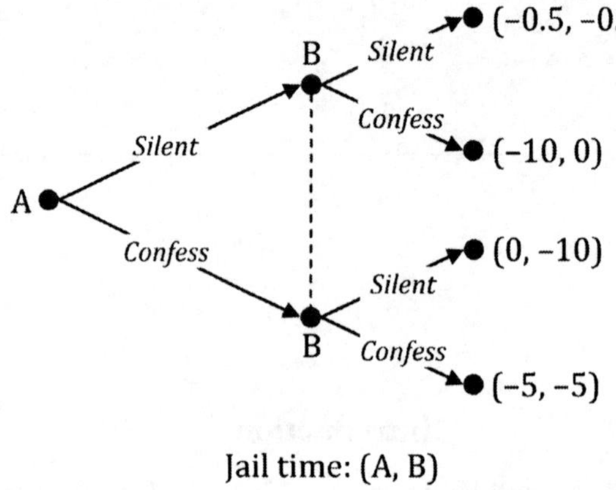

Fig. 2.1.

The extensive-form game depicted in Fig. 2.1 is read from left to right. The closed circle labeled A is referred to as a *decision node*. At this decision node, suspect A decides to *confess* or remain *silent*. These strategy choices are depicted as arrows (called *branches*) that point to decision nodes where B chooses *confess* or *silent*.

In Fig. 2.1, suspect A moves "first" in the sense that he moves without knowing how B will move. Suspect B also moves without knowing how A will move. Player B is said to have **imperfect information** because she does not know whether she is located at the upper or lower decision node. Imperfect information is a feature of all static games. Since B does not know her location in the game tree, her decision nodes are connected with a dashed line. In the lexicon of game theory, B's decision nodes constitute an *information set*. If the player's moves were sequential, B would know her location in the game tree after first observing how A has moved. In this case, B is said to have *perfect information*.

The four closed circles at the far right of Fig. 2.1 are called *terminal nodes*. The entries in parentheses to the right of the terminal nodes summarize the jail time in years for each suspect from each combination of strategies. The first entry is the payoff to A and the second entry is the payoff to B. Since the suspects are assumed rational, larger payoffs (less jail time) are preferred.

The prisoner's dilemma depicted has four possible strategy profiles: {*Confess*, *Confess*}, {*Confess*, *Silent*}, {*Silent*, *Confess*}, {*Silent*, *Silent*}. The worst outcome for A results when he is silent and B confesses. Similarly, the worst outcome for B is when she is silent and A confesses. The best joint outcome is when both are silent since each suspect will receive just 6 months in jail. But, as we will see, both players have a strong incentive to confess, which results in the worst joint outcome.

The Normal Form

A more convenient way to describe the prisoner's dilemma is depicted in Fig. 2.2. This set up, which is referred to as the **normal form**, can be used to illustrate two or three player static games involving a finite number of strategies. The normal form summarizes the players and payoffs from alternative strategy profiles. We will adopt the convention that the first entry in the parentheses is the payoff to the player identified on the left-side of the payoff matrix, and the second entry refers to the payoff to the player identified at the top of the payoff matrix.

		B	
		Silent	*Confess*
A	*Silent*	(−0.5, −0.5)	(−10, 0)
	Confess	(0, −10)	**(−5, −5)**

Jail time: (A, B)

Fig. 2.2.

To understand why each suspect has an incentive to defect, consider the game from A's perspective. If B is silent, it is in A's best interest to confess since this will result in no jail time, compared with six months in jail by not confessing. If B confesses, A's best response is again to confess, since this results in five years in jail compared with ten years by remaining silent. Suspect A's best response is to confess, regardless of the strategy adopted by B. Since the payoff matrix is symmetrical, B will also confess regardless of the strategy adopted by A.

The most likely outcome is for both suspects to confess, even if they agree to remain silent before being separated. Trust is difficult because each suspect has a powerful incentive to confess to get a better deal. Moreover, each suspect wants to avoid the "sucker's payoff" by remaining silent.[3] *Confess* is a strictly-dominant strategy for both suspects. The resulting strategy profile is {*Confess*, *Confess*} with both suspects receiving five-year jail terms.

Coalition games such as the prisoner's dilemma are frequently encountered in business, which explains the widespread use of contracts. A contract is a formal agreement that binds two or more players to a specific course of action in exchange for something of value. Failure of either party to adhere to the terms and conditions of the contract results in legal sanctions. Contracts bind players to agreements by changing the payoffs in the event of defection. The prisoner's dilemma also helps to explain why companies form joint ventures instead of entering into informal strategic alliances.

Nash Equilibrium

The predicted outcome of the game in Fig. 2.2 is referred to as a **Nash equilibrium** in honor of John Forbes Nash, Jr., who with John Harsanyi and Reinhard Selten were awarded the 1994 Nobel Prize in economics for their pioneering work in game theory. A non-cooperative game has a Nash equilibrium when all players adopt a strategy that is the best response to the strategies adopted by the other players, and neither player can get a better payoff by unilaterally switching strategies.[4]

Nash revolutionized the economics profession when he first proposed the idea of a "fixed-point equilibrium" in his twenty-seven-page doctoral dissertation published in 1950. This concept seemed to contradict Adam Smith's famous metaphor of the "invisible hand," which suggested that the welfare of society is maximized when individuals pursue their private interests. As we saw in Fig. 2.2, it was in the suspects' best interest to cooperate.

Nash's epiphany was dramatized in the film *A Beautiful Mind*, which won the 2001 Academy Award for Best Picture. The pivotal scene in this movie, which was loosely based on the best-selling biography written by Sylvia Nasar (1998), takes place when five beautiful women walk into a tavern, one of whom is a stunning blonde. Nash and his friends plot ways for winning a date with the blonde. In this fictitious scene, Nash reasons that if everyone goes for the blonde, they will block each other. Moreover, they will be rejected by the brunettes who are insulted at being second choices. Nash concludes that the optimal strategy is for everyone to ignore the blonde and go for the brunettes. While this strategy results in no one getting the blonde, at least everyone gets a date. Cooperation—not competition—results in the best outcome.

The concept of a Nash equilibrium is powerful because it exists for all non-cooperative games, although not always in pure (nonrandom) strategies. Many pure-strategy static games have multiple Nash equilibria, and some have none at all. On the other hand, all static games have a unique Nash equilibrium when a finite number of pure strategies are randomized. Because of this, the concept of a Nash equilibrium is virtually unchallenged as a solution concept to situations involving non-cooperative strategic behavior.

The prisoners' dilemma provides important insights into the strategic behavior of firms with market power. Consider an industry consisting of just two firms (duopoly). Suppose that firms A and B are deciding whether to charge a high or low price for their products. Each firm's profit depends on its price and the price charged by the other firm. If A charges a high price, for example, B can increase its sales at A's expense by charging a low price, and *vice versa*. On the other hand, both firms can share monopoly profits by forming a price-fixing coalition.

Consider, for example, the pricing game depicted in Fig. 2.3 in which profits are in millions of dollars and larger payoffs are preferred. In this game, each firm has a strictly-dominant *low-price* strategy. If A charges a high price, B's profit is $1.5 million by charging a low price, compared with a profit of $1 million by charging a high price. If A charges a low price, B's profit is $100,000 if it charges a high price and $250,000 if it charges a low price. B's best response is to charge a low price, regardless of the strategy adopted by firm A. Firm B has a strictly-dominant *low-price* strategy. Since the payoffs in Fig. 2.3 are symmetrical, A has a strictly-dominant *low-price* strategy as well. The resulting payoffs when both firms adopt their strictly-dominant strategies are highlighted in boldface.

		B	
		High price	*Low price*
A	*High price*	(1.0, 1.0)	(0.1, **1.5**)
	Low price	(**1.5**, 0.1)	**(0.25, 0.25)**

Profits: (A, B)

Fig. 2.3.

The strategy profile {*Low price*, *Low price*} in Fig. 2.3 is a Nash equilibrium because charging a low price is the best response to the strategy adopted by the other firm. Moreover, at this strategy profile neither firm can improve its payoff by unilaterally switching strategies. What is interesting about this outcome is that the pursuit of individual self-interest results in a less-than-optimal outcome for both firms. By forming a coalition and charging a high price, each firm can earn $1 million, compared with $250,000 by not cooperating. In the absence of an effective enforcement mechanism that binds the firms to this agreement, however, this coalition will likely break down. Each firm can improve its payoffs and earn $1.5 million by unilaterally charging a low price.

The coalition game depicted in Fig. 2.3 is revealing because it shows how competition can lower a firm's value, and how cooperation can create profit opportunities. In fact, the main reason for studying the prisoner's-dilemma-type games is to avoid suboptimal outcomes that result from rivalrous behavior. The implications of this for such problems as environmental pollution and traffic congestion suggest that in some cases society may be made better off when decision makers cooperate to achieve a mutually advantageous outcome. Even when collusion is legal, cooperation among firms has often proven to be highly unstable since there is a strong incentive to cheat, perhaps because there is a fundamental distrust of a competitor's motives and intentions. In such cases, an enforcement mechanism that penalizes defectors by changing the payoffs is required to bind the coalition.

Shortcut for Finding Pure-Strategy Nash Equilibria

An efficient way to identify a Nash equilibrium strategy profile in static games in normal form is by underlining the highest payoff associated with each player's best response. Consider, for example, the pricing game depicted in Fig. 2.3. Firm A's best response to a high-price strategy by firm B is to charge a low price because $1.5 million is greater than $1 million. Underline this highest payoff and repeat this procedure for both players until all strategy profiles have been examined. To determine if a pure-strategy Nash equilibrium strategy profile exists, look for the cell in the payoff matrix in which both payoffs are underlined. The only cell in Fig. 2.3 for which the payoffs of both players are underlined is the strategy profile {*Low price*, *Low price*}. In much of the remainder of this text, best-response payoffs will be highlighted in boldface.

The reader is cautioned that this technique often suggests the existence of multiple Nash equilibria, or none at all. When this occurs, a closer examination of the strategic interaction of the players is warranted to determine each player's best-response to the anticipated moves of the other players.

Determinants of Business Collusion

In coalition games it is in the best interest of players to cooperate, but each player has an incentive to defect. Coalitions will fail unless the agreement includes some mechanism that punishes defectors, such as a contract. Factors that might account for the emergence of a business coalition include the number of firms with similar interests, the size of the firms relative to the industry as a whole, and the visibility of the agreement.

Number of Firms with Similar Interests

Collusion is more likely among firms with similar interests, but this is no guarantee of success. Collusion is difficult when there are a large number of firms. This is because it is increasingly difficult and costly to monitor compliance as the number of member firms increases. If there are n parties to the agreement, each member must monitor the behavior of the other $n - 1$ members. The total number of monitoring arrangements necessary to police the cartel is $n(n - 1)$. Sanctions become increasingly more difficult to impose as membership in the cartel increases, which increases the possibility that the cartel will break down. A two-firm coalition requires just $2(2 - 1) = 2$ monitoring arrangements. The Vienna-based Organization of Petroleum Exporting Countries (OPEC) sets production quotas for its 13 members (Algeria, Angola, Ecuador, Indonesia, Iran, Iraq, Kuwait, Libya, Nigeria, Qatar, Saudi Arabia, United Arab Emirates, and Venezuela). This cartel requires $13(13 - 1) = 156$ arrangements to monitor compliance! Policing OPEC production-sharing agreements is made even more difficult because members have widely diverse cultural, historical, economic, and political traditions. This helps to explain why OPEC oil ministers frequently hammer out new production-sharing agreements. The incentive to cheat, especially by members with low production quotas, is very strong.

Firm Size Relative to the Industry

Economies of scale exist in the monitoring and policing of collusive agreements. It is less expensive for a large firm to monitor the behavior of a few large rivals than it is for a small firm to monitor the behavior of many small rivals.

Visibility

Collusion may be explicit or tacit. A cartel is explicit if its members meet overtly to hammer out an agreement. An explicit collusive agreement openly specifies each player's responsibilities, such as production quotas, pricing, advertising, market shares, sanctions, and so forth.

In many countries, business collusion is illegal. This is not to say that firms do not practice tacit collusion, as seems to occur when airlines raise ticket prices, even though each airline has an incentive to slash prices to increase sales. Tacit collusion often results from shared understanding of the business environment or by observing how rivals behaved in the past. Unlike legal coalitions, the use of contracts to bind players to an agreement is not possible.

There are several interesting techniques used by firms to discourage rivals from upsetting the *status quo*. In retailing, a firm may adopt a "we will not be undersold" policy or provide low-price guarantees. Firms have less incentive to violate implicit price agreements when they know that any price reduction will be matched by rivals.

Another method for fixing prices is a most-favored-customer policy. This policy guarantees that a customer will get the best price offered to any other customer. If another customer were to receive a lower price, all most-favored customers will also receive the same low price. A firm that adopts a most-favored-customer policy is better able to withstand pressures to lower prices for anyone.

Price leadership is when the pricing decisions of one firm are mimicked by the rest of industry. Suppose that Home Depot is the industry's price leader. When an industry has a recognized price leader, price wars are avoided and all firms benefit from the higher price.

The enforcement mechanisms outlined above are only effective if the threat of retaliation is credible. Coalitions will fail if players are unwilling or unable to punish defectors. Collusive arrangements are more likely to endure if the threat of punishment is sufficient and certain. The cost of punishing defectors, however, can be quite high. Firms must be prepared to lower the price charged not to the defector's customers, but their own customers as well. The rogue firm must believe that defection will fail.

Solved Exercise

Home Depot and Lowe's Home Improvement are do-it-yourself superstores. Suppose that each firm is considering lowering the price of a popular hardware item. While both companies have a core of loyal customers, there is a third group that is very price sensitive. The expected payoffs from alternative strategy profiles are summarized in Fig. 2.4.

a. *If larger payoffs are preferred, does either player have a strictly-dominant strategy?*
b. *Does this game have a Nash equilibrium?*
c. *What can the firms do to stabilize an illegal price-fixing agreement?*

		Loew's	
		Lower price	*Higher price*
Home Depot	*Lower price*	**(90, 80)**	**(110,** 50)
	Higher price	(60, **110)**	(100, 90)

Profits: (Home Depot, Loews)

Fig. 2.4.

Solution

a. *The reader should verify that it is in each player's best interest to charge a* lower *price regardless of the pricing strategy adopted by its rival.*
b. *Since both players should adopt its strictly-dominant* lower *price strategy, the Nash equilibrium strategy profile is {Lower price, Lower price}.*
c. *A tacit price-fixing coalition may emerge since it is in both players' best interest to charge a* higher *price. The strategy profile {Higher price, Higher price}, however, is not stable since both firms have an incentive to defect. The firms may be able to stabilize the coalition by offering low-price guarantees, instituting a most-favored-customer policy, or pledging that they will not be undersold. In this way, the firms serve notice that defection will not go unpunished.*

Practice Exercises

2.1 What is a coalition game?

2.2 What is a strictly-dominant strategy?

2.3 Describe completely what is meant by a Nash equilibrium strategy profile?

2.4 The prisoner's dilemma is an example of a coalition game. Since cooperation results in the best outcome for each suspect, the resulting strategy profile must be a Nash equilibrium. Do you agree?

2.5 The existence of a Nash equilibrium confirms Adam Smith's famous metaphor of the invisible hand. Do you agree with this statement? If not, why not?

2.6 In the country of Arcadia two equally sized companies, Auburn Motorcar Company and Cord Automobile Corporation, dominate the domestic automobile market. Each company can produce 500 or 750 midsized automobiles a month. The four strategy profiles for this game are {*750 cars, 750 cars*}, {*750 cars, 500 cars*}, {*500 cars, 750 cars*}, and {*500 cars, 500 cars*}. The profits for (Auburn, Cord) are (5, 5), (3, 6), (6, 3), and (4, 4), respectively.

a. If larger payoffs are preferred, does either firm have a strictly-dominant strategy?
b. What is the Nash equilibrium strategy profile for this game?

c. Is it in the best interests of the players in this game to form a coalition? Is this coalition stable?

2.7 Suppose that Tsunami Corporation and Cyclone Company are considering a change in their pricing policies. The four strategy profiles for this game are {*High price*, *High price*}, {*High price*, *Low price*}, {*Low price*, *High price*}, and {*Low price*, *Low price*}. The respective payoffs for (Tsunami, Cyclone) in millions of dollars are (6, 6), (4, 8), (8, 4), and (5.5, 5.5).

a. If larger payoffs are preferred, does either firm have a strictly-dominant strategy?

b. Does this game have a Nash equilibrium strategy profile?

c. Is it in the best interests of the players in this game to form a coalition? Is this coalition stable?

2.8 The Global Mining and Recovery (GLOMAR) Corporation and the Petroleum Exploration (PETROX) Company own leases on public land that lies directly above a crude oil deposit. Both companies must decide whether to drill a high-volume or a low-volume well. If GLOMAR and PETROX drill high-volume wells, each earns a profit of $24 million. If each company drills a low-volume well, each earns a profit of $52 million. If one company drills a high-volume well and the other drills a low-volume well, the company drilling the high-volume well will earn $70 million and the company drilling the low-volume well will earn $12 million.

a. If larger payoffs are preferred, does either firm have a strictly-dominant strategy?

b. Does this game have a Nash equilibrium strategy profile?

c. Is it in the best interests of the players in this game to form a coalition? Is this coalition stable?

2.9 Why do fast-food restaurants tend to cluster in the same location? Suppose that two fast-food franchises, A and B, initially locate near each other in the center of town. The state recently opened a new exit ramp from the nearby interstate highway. Although both restaurants are operating profitably, the restaurant owners calculate that by relocating near the interstate highway they will continue to receive some in-town business, but will benefit from sales to drivers using the exit as a rest stop. There are four possible strategy profiles: {Exit ramp, Exit ramp}, {Exit ramp, Downtown}, {Downtown, Exit ramp}, and {Downtown, Downtown}. The respective payoffs for (A, B), which are in millions of dollars are: (0.15, 0.15), (1.0, 0.1), (0.1, 1.0), and (0.25, 0.15).

a. Does either player have a strictly-dominant strategy?

b. Does this game have a Nash equilibrium strategy profile?

c. Are these restaurants better off by forming a coalition? Is this coalition stable?

Notes

1. This important contribution to the development of game theory did not first appear in a scholarly journal, but in the classroom. In 1950, the well-known mathematician Albert J. Tucker was a visiting professor at Stanford University. While attempting to illustrate to a group of psychologists the difficulty of analyzing certain kinds of games, Tucker developed the prisoner's dilemma. S. J. Hagenmayer (1995) wrote that "Mr. Tucker's simple explanation has since given rise to a vast body of literature in subjects as diverse as philosophy, ethics, biology, sociology, political science, economics, and, of course, game theory."

2. Variations of the prisoner's dilemma can be found in Dixit and Nalebuff (1991), Schotter (1985), and Luce and Raiffa (1957), and elsewhere.

3. A large number of experiments have confirmed this prediction. In experiments conducted by Cooper, DeJong, Forsythe and Ross (1996), test subjects were asked to play the prisoner's dilemma twenty times against different anonymous rivals. As the players gained experience, the strategy profiles of the players converged toward the predicted Nash equilibrium. In the first five rounds, for example, rivals cooperated 43 percent of the time. This cooperation fell to just 20 percent in the last five rounds.

4. Formally, suppose a player's payoff function u_i is defined over the set of strategies S, where $u_j(s)$ is player j's payoff for a given strategy $s_i \in S_i$. The strategy profile $\{u_A^*(u_B^*), u_B^*(u_A^*)\}$ in a one-time, two-player (j = A, B), static game is a Nash equilibrium when strategy $u_A^*(s)$ is player A's best response to strategy $u_B^*(s)$ adopted by player B, strategy $u_B^*(s)$ is player B's best response to strategy $u_A^*(s)$ adopted by player A.

CHAPTER THREE

Strategic Moves and Deterring Defection

Introduction

In the previous chapter we saw that coalitions may fail if the players have an incentive to defect. To prevent this, an enforcement mechanism that punishes defectors may be necessary. Enforcement mechanisms work either by eliminating the incentive to defect, or by making it difficult to back out once a commitment has been made.

Strategic Moves

An enforcement mechanism to prevent defection is an example of a **strategic move**, which is an observable and irreversible move designed to alter the behavior of a rival. Strategic moves can take the form of threats or promises. For a strategic move to succeed, it must be credible. Rivals must believe that a player is ready to stand by its commitments.

A **threat** is a countermove intended to punish rivals who do not cooperate. There are two types of threats. The purpose of a *compellent threat* is to compel a rival to behave in a way that is in the player's best interest. A *deterrent threat* is meant to deter a rival from behaving in a way that is not in the player's best interest. In each case, both players are made worse off if the threat is carried out. By contrast, a **promise** is a countermove that rewards a rival for cooperating. As with threats, there are compellent and deterrent promises. In both instances, the player making the promise has an incentive to break his or her word.

Deterring Defection

This section discusses eight possible methods for deterring defection in coalition games. The first two (contracts and reputation) work by changing the payoffs in the event of defection. The remaining methods make it difficult for players to back out of an agreement.

Contracts

There are many business situations in which it is in the firms' best interest to form a coalition, which ultimately breaks down because the players had a unilateral incentive to defect. For this reason, contracts are a common feature of many business relationships. A **contract** is a formal agreement that obligates the parties to perform, or refrain from performing, a specified act in exchange for something of value. In the event of a **breach of contract**, the courts can order restitution and impose punitive damages. The resulting **breach remedy** strengthens a coalition by removing the incentive to defect.

Breach remedies must be sufficient enough to ensure that cooperation results in a better payoff than defection. Even a sufficiently punitive breach remedy may not be enough to guarantee compliance. It must not be possible to renegotiate a contract since this will weaken a rival's commitment to the coalition.[1]

For a contract to effectively deter defection it must be enforceable and the enforcement agent must be properly motivated. There is a law on the books in Pennsylvania, for example, that requires motorists to disassemble their car and hide the pieces under the nearest bush at the sight of an oncoming team of horses. While this contract is technically enforceable, it is not in the best interest of the state or its citizens to do so.

A contract with a legally enforceable breach remedy is an effective method of enforcing compliance in coalition games because it makes unilateral defection unprofitable. When it is difficult, or impossible, to enforce a contract, however, some other mechanism to deter defection may be required.[2] One such powerful incentive is a player's reputation.

Reputation

In business, a firm's reputation can be critical to a firm's success. Reputation may not mean too much in a game that is played only once, since defection does not affect future game play. This helps explain the disreputable behavior of used-car salespeople and telemarketers. On the other hand, a player's credibility will be tested in games that are played repeatedly. Establishing and maintaining a reputation as someone who follows through with threats, honors commitments, and keeps promises can be a valuable strategic asset.

Countries such as the United States and Israel have an official policy of not negotiating with terrorists. The rationale behind this policy, and the importance of standing by this commitment, is straightforward. Terrorists must believe that their actions will not yield benefits. Acquiescing to demands will only encourage more terrorism and should be steadfastly rebuffed.

Paradoxically, breaking your word can sometimes lend credibility to a commitment. This is because it may rule out future actions that may not be in a rival's best interest. Suppose, for example, that terrorists highjack an airplane and threaten to kill the passengers unless their demands are met. The terrorists might reason that officials have no choice but to negotiate since innocent lives are at stake.

Is there a way out of this dilemma? One possibility is for the government to agree to the terrorists' demands. Once the hostages are released, however, the authorities attack the terrorists. Having been double-crossed once before, terrorists will think twice about the credibility of government assurances. As the saying goes, "once burned, twice shy." By breaking its commitment not to negotiate, the government's no-negotiation pledge actually becomes more credible.

Establishing and maintaining a player's reputation to enhance credibility has important strategic implications. On the other hand, establishing a reputation for being irrational can have the same effect. Threats that are implausible to a rational person may have strategic value when uttered by a sociopath. When a lunatic makes a threat, no matter how off the wall, it must be taken seriously. This is the dilemma that often confronts U.S. government officials when dealing with rogue states ruled by political and religious fanatics seeking nuclear weapons, or parents dealing with unruly young children. Alternatively, reasonable promises by psychopaths, such as the non-aggression pledges made by Adolf Hitler following territorial concessions in the run up to World War II, must be weighed with a healthy dose of skepticism.

Cutting off Communications

Contracts and reputation strengthen coalitions by removing the incentive to defect. In some cases, it may be possible to defer defection by closing the door to future communications, thereby making a player's commitment irreversible. An extreme example of this is a Last Will and Testament. Promises made in life are irreversible in death. Sending a registered letter is another example of an irreversible commitment since it cannot be retrieved. Moreover, neither the sender nor the recipient can claim ignorance of the letter's contents.

One problem with cutting off communications is that it becomes difficult to monitor compliance. For this reason, it may be necessary to employ the services of a third party to ensure compliance. This is the role of an estate executor in a Last Will and Testament.

Preventing Retreat

Cutting off escape routes can strengthen coalitions by making retreat impossible. Edward Land, founder of Polaroid, refused to diversify out of the instant photography business. As a result, rivals such as Kodak were well aware of Land's messianic commitment to defend his company's territory against patent infringement at all costs. Walter Mondale knew that he would irreversibly alienate some voters in his 1984 presidential bid by pledging to raise taxes to stanch the flow of red ink in the federal budget. He hoped to cement his position among voters who favored his economic program. In 1989, Egon Krenz, the prime minister of the German Democratic Republic, dismantled portions of the Berlin Wall to lend credibility to his proposed economic reform program. This action worked because it denied the East German government the opportunity to retreat, lest they run the risk of a mass exodus to the West.

Burning bridges to bolster a player's commitment was discussed in a scene from the blockbuster movie *The Hunt for Red October,* which was based on Tom Clancy's best-selling novel of the same name. Captain Marco Ramius (portrayed by Sean Connery) and a group of Soviet naval officers conspire to defect and deliver the nuclear submarine *Red October* into the hands of the U.S. Navy. Ramius is criticized by a co-conspirator for his decision to inform Soviet Admiral Yuri Padorin (Peter Zinner) of his intentions, which prompted a frantic search and destroy operation. In explaining his decision, Ramius explained that in 1519 Hernando Cortés landed in Mexico with a force of just six hundred men, twenty horses, and ten small cannons to conquer an Aztec nation of more than five million people. "When Cortés landed in the New World he burned his ships. As a result, his men were well motivated." By denying his men the opportunity to retreat, Cortés gave them a stark choice: fight or die.

Brinksmanship

Thomas Schelling (1960, p. 200) describes brinksmanship as "the deliberate creation of a recognizable risk . . . that cannot be completely controlled." Dixit and Nalebuff (1991, p. 155) refer to brinksmanship as "leaving the outcome beyond your control." Brinksmanship lends credibility to a commitment by exposing a rival to a shared risk. An opponent who upsets the *status quo* may result in both players slipping over the brink "whether we want to or not, carrying him with us" (Schelling 1960, p. 200).

An example of brinksmanship was the Cuban missile crisis, which brought the United States and the Soviet Union to the brink of nuclear war in 1962. The crisis began when President John F. Kennedy committed a deliberate act of war by blocking Soviet ships carrying nuclear-tipped missiles to Cuba. Fortunately, the crisis was defused when the Russians agreed to remove missiles from Cuba in exchange for a promise by the president not to invade Cuba and to remove U.S. missiles from Turkey.

For brinksmanship to be effective it must be measured since accidents or miscalculations could cause a strategic situation to spiral out of control. An example of this can be found in Eugene Burdick and Harvey Wheeler's 1962 best-selling novel and underappreciated 1964 movie, *Fail-Safe*. A combination of human and computer error resulted in an American strategic bomber being ordered to drop its nuclear payload on Moscow. The U.S. president and military wrestle with the crisis, but despite their best efforts, the bomber breaches Soviet defenses to complete its mission.

Fearing an all-out nuclear confrontation, the American president (portrayed by Henry Fonda) offers to drop a nuclear bomb on New York City to avoid a retaliatory strike that will escalate into a full-scale, mutually-destructive nuclear war. The Soviet premier accepts the offer and nuclear Armageddon is avoided. *Fail-Safe* highlights the benefits and dangers of brinksmanship. Although this policy lends credibility to a commitment, small errors can spiral out of control resulting in a less-than-optimal outcome.

Incrementalism

When the benefits of unilateral defection are substantial, forming a viable coalition is difficult. In this case, it may be possible to reach an accommodation by deconstructing the process into a series of smaller transactions. Forming lesser coalitions with smaller payoffs may lead to an outcome in which the whole may be greater than the sum of its parts. Players are more willing to form coalitions when the sucker's payoff is small. Once credibility is established, the tactic is repeated. As trust is established and reinforced, the players may be more willing to make larger commitments.

The incremental approach works by breaking the process down into a series of less risky games. As an example, suppose a homeowner hires a contractor to rebuild his front porch. Unless both players have had prior dealings, neither will completely trust the other. The homeowner will probably not pay the full contract price up front for fear that the contractor will not use quality materials, do first-rate work, or complete the job in a timely manner. On the other side, the contractor may not completely trust the homeowner to pay once the job is finished.

To overcome this lack of trust, the homeowner might agree to pay the contractor on a periodic basis. In this way, the homeowner can monitor the contractor's progress, while the contractor's risk of nonpayment is reduced. Satisfaction by both parties also paves the way to future cooperation. The contractor's business may also benefit from an enhanced reputation and satisfied customer referrals.

A potential problem with the incremental approach is the *end-of-game problem*. If either player believes the other will defect in the last stage of the game because there is no possibility to retaliate, previously established trust may completely unravel. Suppose the homeowner believes that the contractor will not complete the job unless he receives a premium over the agreed-upon price. This tactic may be successful, especially if the cost of locating another contractor to complete the job is sufficiently high.

To avoid being held up in the final round, the homeowner may consider cancelling the contract before the final round and bringing in another contractor to complete the job. This transforms the penultimate round into the last round. Since both parties will project forward and reason back, the contractor may end the relationship in the next-to-penultimate round, and so on until the entire relationship disintegrates. One way to avoid the end-of-game problem is to extend the relationship beyond the project's completion, such as by providing warranties and money-back guarantees.

Teamwork

A player with a strong incentive to defect may draw strength from others. Teamwork and peer pressure can be powerful inducements, especially when a player values the opinion of others. In the midst of the horror and chaos of battle, for example, a lone soldier may be tempted to desert her post. But, if the soldier is a member of a squad, platoon, company, or battalion, then "duty, honor, and country" can result in a strong personal commitment to a well-defined objective since desertion can lead to panic, which could result in a rout and widespread loss of life. For this reason, a soldier's military training is supplemented with credible assurances of death by firing squad in the event of desertion. The possibility of death resulting from an attack becomes more attractive than the certainty of death from desertion.

Agents

Another method of reinforcing a commitment is to use negotiating agents who lack the authority to compromise or make concessions. This last method helps explain why professional athletes and entertainment celebrities willingly pay 10 percent agency commission to negotiate service contracts on their behalf. It also explains the role of union leadership in collective bargaining sessions with management. In both cases, the principals' agents can be dismissed if they exceed their negotiating mandate or do not act in their clients' best interests.

Practice Exercises

3.1 What is the difference between a compellent and deterrent threats and promises?

3.2 Explain how contracts and reputation differ from cutting off communications, preventing retreat, incrementalism, teamwork, and agents making threats credible.

3.3 An archaic Pennsylvania state law requires motorists in rural areas to stop every mile and launch a signal rocket. Not only do drivers routinely violate this law, they would probably be arrested if they did. There are many such laws currently on the books throughout the United States and around the world. What does this imply about credible commitments and the rule of law in democratic societies?

3.4 Consider the one-time, static game involving players A and B. There are four strategy profiles: {*A1*, *B1*}, {*A1*, *B2*}, {*A2*, *B1*}, and {*A2*, *B2*}. The respective payoffs for (A, B) are (3, 3), (1, 10), (10, 1), and (2, 2).

a. What strategy profile results in the best payoff for both players?
b. Is the strategy profile identified above a pure-strategy Nash equilibrium?
c. Suppose that the players sign a legally binding contract that commits the parties to a strategy that results in the best payoff. The specified remedy in the event of a breach is two times the contract payoff. Is this breach remedy sufficient to bind the players to the coalition?
d. Would a breach remedy of three times the payoff under the contract be sufficient to prevent defection?

Notes

1. Thomas Schelling (1989) cites the example of a drug rehabilitation clinic in Denver that caters to wealthy clients. Patients are required to sign a letter of self-incrimination that attests to their substance abuse and addiction as a precondition for admission. It is understood that the letter will be made public if the patient fails a random drug test while at the clinic. Schelling noted that many patients subsequently tried to buy their way out of this agreement, but the clinic had an incentive to follow through with this threat since its reputation was at stake.

2. Hollywood mogul Sam Goldwyn once quipped that a "... verbal contract is not worth the paper it's written on" (Berg 1989, p. 386).

CHAPTER FOUR

Competition Games

Introduction

The prisoner's dilemma in Fig. 2.2 is an example of a *coalition game* in which it is in the players' best interest to cooperate, but where each player has an incentive to defect. In a one-time, two-player, static **competition game** the players rank payoffs from alternative strategy profiles in reverse order.[1] As one player's payoff increases, the other player's payoff decreases. In this chapter, we will consider competition games in which either or both players have a dominant strategy.

Strictly-Dominant Strategies

Consider the game in Fig. 4.1. If B moves *B1*, A's best response is *A1* because it results in a better payoff. If B moves *B2*, A again moves *A1* for the same reason. Since A's best response is *A1* regardless of the strategy adopted by B, *A1* is said to be a *strictly-dominant strategy*. By contrast, B's best response varies with the strategy adopted by A. If A moves *A1*, B's best response is *B1*. If A moves *A2*, B's best response is *B2*.

		B	
		B1	*B2*
A	*A1*	**(100, 200)**	(**150**, 75)
	A2	(50, 50)	(120, **150**)

Payoffs: (A, B)

Fig. 4.1.

Although B does not have a strictly-dominant strategy, B expects A to play his strictly-dominant strategy. In a one-time, two-player, static game it is only necessary for one rational player to have a dominant strategy for a Nash equilibrium to exist. In Fig. 4.1, the strategy profile {*A1*, *B1*} is a Nash equilibrium since each player's move is a best response, and neither player can improve his or her payoff by unilaterally switching strategies. Note that in this game there is no coalition that will make both players better off.

Having a strictly-dominant strategy does not guarantee the best possible payoff. In Fig. 4.1, B does not have a strictly-dominant strategy, but her best response results in the best possible payoff. Player A does not receive the best possible payoff, even though he does have a strictly-dominant strategy.

Weakly-Dominant Strategies

In Fig. 4.2, which is a variation of the game in Fig. 4.1, A does not have a strictly-dominant strategy. This is because if B moves *B1*, A's best response is *A1*. But, if B moves *B2*, A is indifferent between *A1* and *A2*. *A1* is a **weakly-dominant strategy** because *A1* and *A2* will result in the same payoff if B moves *B2*, but *A1* will result in a strictly better payoff if B moves *B1*. For this reason, *A1* is a best response to both *B1* and *B2*.

		B	
		B1	*B2*
A	*A1*	**(100, 200)**	(**120**, 75)
	A2	(50, 50)	(120, **150**)

Payoffs: (A, B)

Fig. 4.2.

In the static game depicted in Fig. 4.2, the strategy profile {*A1*, *B1*} is a Nash equilibrium since it represents a best response by both players. Since strictly- and weakly-dominant strategies are best responses, they are collectively referred to as **dominant strategies**.

Solved Exercise

Firm X is considering entering a market in which firm Y is a monopolist. Firm Y can adopt a predatory *or* competitive *pricing strategy. The present values of profits in millions of dollars are summarized in Fig. 4.3.*

a. *If larger payoffs are preferred, does either firm have a dominant strategy?*
b. *Does this game have a Nash equilibrium?*

		Y	
		Predatory	*Competitive*
X	*Enter*	(−2.5, 1)	**(3, 4)**
	Stay out	(**0**, 7)	(0, **7**)

Profits: (X, Y)

Fig. 4.3.

Solution

a. *X does not have a dominant strategy. If Y moves* predatory, *X's best response is to* stay out. *If Y moves competitive, X's best response is to enter. X's best response changes with the strategy adopted by Y. By contrast, Y has a weakly-dominant* competitive *pricing strategy. Y is indifferent between* predatory *and* competitive *if X stays out. On the other hand, if the move by X is to enter, Y's best response is competitive.*
b. *Y's best response to any move by X is a* competitive *pricing strategy. X's best response is to* enter. *The Nash equilibrium strategy profile for this game is {Enter, Competitive} since neither firm can improve its payoff by unilaterally switching strategies.*

Iterated Elimination of Dominated Strategies

So far, we have considered one-time, static games involving two players and two strategies. Suppose, however, that two players have three or more strategies. In games where at least one player has a dominant strategy, identifying a Nash equilibrium (if one exists) is reasonably straightforward. If neither player has a dominant strategy, it may still be possible to identify a Nash equilibrium by iteratively eliminating strictly-dominated strategies. A **strictly-dominated strategy** is never chosen, regardless of the strategies adopted by other players.

To illustrate this process, consider the static game in Fig. 4.4. Regardless of the strategy adopted by B, A will never move A0 since the resulting payoffs are strictly worse than the payoffs from either of the remaining two strategies. This strategy should be eliminated from further consideration. Note also that regardless of the

		B		
		B0	*B1*	*B2*
	A0	(0, 0)	(0, 200)	(0, 250)
A	*A1*	(200, 0)	(100, 200)	(150, 75)
	A2	(250, 0)	(50, 50)	(120, 150)

Profits: (A, B)

Fig. 4.4.

strategy adopted by A, *B0* should also be eliminated for the same reason. After eliminating *A0* and *B0* we are left with the static game depicted in Fig. 4.1. But, this is not the end of the story.

In Fig. 4.1, regardless of the strategy adopted by B, *A2* is strictly dominated by *A1* and should also be eliminated. The remaining game involves one strategy for A (*A1*) and two strategies for B (*B1* and *B2*). Since *B2* is strictly dominated by *B1*, *B2* should be eliminated. What remains is the Nash equilibrium strategy profile {*A1*, *B1*}, which was first identified by identifying the players' strictly-dominant strategies.

The rationale behind eliminating strictly-dominated strategies may be summarized in Sherlock Holmes's famous dictum: "When you have eliminated the impossible, whatever remains, however improbable, must be the truth" (Doyle, 1890, Ch. 6). Actually, this assertion by Sir Arthur Conan Doyle's super sleuth is an exaggeration. Eliminating dominated strategies does not guarantee that a Nash equilibrium (if one exists) will emerge. If a player has a strictly-dominant strategy, the order in which strictly-dominated strategies are eliminated is irrelevant. On the other hand, the order in which *weakly-dominated* strategies are removed could affect the outcome of the search.

Solved Exercise

If larger payoffs are preferred, identify the Nash equilibrium strategy profile (if it exists) for the game depicted in Fig. 4.5 by eliminating strictly-dominated strategies.

		B			
		B1	*B2*	*B3*	*B4*
A	*A1*	(1, 0)	(2, 3)	(**4**, 0)	(1, **4**)
	A2	(2, 1)	(**4**, **5**)	(3, 1)	(3, 4)
	A3	(2, **4**)	(1, 3)	(2, 0)	(0, 1)
	A4	(**3**, 2)	(3, 4)	(2, 1)	(**5**, 3)

Payoffs: (A, B)

Fig. 4.5.

Solution

The reader should verify that neither player has a strictly-dominant strategy. On the other hand, A3 is strictly-dominated (none of the payoffs in that row are boldfaced) by the other strategies and should be eliminated. Similarly, B3 is strictly dominated by the other strategies and should also be eliminated.

In the reduced normal-form game in Fig. 4.6, A1 is strictly dominated by A2 and A4. B1 is strictly dominated by B2 and B4. After these strictly-dominated strategies are eliminated the resulting normal-form game is depicted in Fig. 4.7.

In Fig. 4.7, B4 is strictly dominated by B2. After B4 is eliminated, A4 is strictly dominated by A2 and should also be eliminated. The remaining strategy profile {A2, B2} is a unique Nash equilibrium.

		B		
		B1	*B2*	*B4*
	A1	(1, 0)	(2, 3)	(1, **4**)
A	*A2*	(2, 1)	(**4**, **5**)	(3, 4)
	A4	(**3**, 2)	(3, **4**)	(**5**, 3)

Payoffs: (A, B)

Fig. 4.6.

		B	
		B2	*B4*
A	*A2*	(**4**, **5**)	(3, 4)
	A4	(3, **4**)	(**5**, 3)

Payoffs: (A, B)

Fig. 4.7.

Three-Player Games

Analyzing pure-strategy static games involving two players and a finite number of pure strategies is reasonably straightforward. The process becomes exponentially more complicated, however, as the number of players increases. This section describes the process of searching for a Nash equilibrium in pure-strategy static games involving three players. The search in games involving more than three players requires the use of more advanced mathematical techniques and is beyond the scope of this book.

Fig. 4.8 summarizes the strategies and payoffs for a static game involving three players, each having two strategies. The search begins by treating the right and left sides of Fig. 4.8 as stand-alone static games involving players A and B. The payoff matrix on the left is a static game between A and B when C moves *C1*. The payoff

		C			
		C1		*C2*	
		B		B	
		B1	*B2*	*B1*	*B2*
A	*A1*	(**25**, **25**, **25**)	(12, **24**, **12**)	(**12**, **12**, 24)	(**4**, 8, 8)
	A2	(24, **12**, **12**)	(8, 8, 4)	(8, **4**, 8)	(3, 3, 3)

Profits: (A, B, C)

Fig. 4.8.

matrix on the right is a static game between A and B when C moves *C2*. In each of these sub-games, A has a strictly-dominant *A1* strategy and B has a strictly-dominant *B1* strategy.

The next step is to find C's best response by comparing each combination of moves by A and B. Consider, for example, the strategy profile {*A1*, *B1*, *C1*} on the left and the strategy profile {*A1*, *B1*, *C2*} on the right. The best response by C is *C1* since 25 > 24. Similarly, comparing strategy profile {*A2*, *B2*, *C1*} on the left and the strategy profile {*A2*, *B2*, *C2*} on the right, C's best response is *C1* since 4 > 3. Proceeding in this fashion, the Nash equilibrium strategy profile for this game is {*A1*, *B1*, *C1*}.

Non-Dominant Strategies

In the games thus far considered, either or both players had a dominant strategy, which resulted in a unique Nash equilibrium. But, what if neither player has a dominant strategy? Is it still possible for a static game to have a unique Nash equilibrium?

In the game depicted in Fig. 4.9, neither player has a dominant strategy. The best response by each player alternates with the strategy adopted by the other player. If A moves *A1*, B's best response is *B1*. If A moves *A2*, B's best response is *B2*. The same can also be said of A's best responses to the moves of B. Using the search algorithm described in chapter 2, this game has two candidates for Nash equilibria: {*A1*, *B1*} and {*A2*, *B2*}. Unfortunately, it is not possible to identify the most likely outcome based on the information provided since A prefers {*A2*, *B2*} and B prefers {*A1*, *B1*}.

		B	
		B1	*B2*
A	*A1*	**(100, 200)**	(120, 75)
	A2	(50, 50)	**(150, 150)**

Profits: (A, B)

Fig. 4.9.

As the game depicted in Fig. 4.9 demonstrates, not all static games in pure strategies have a unique Nash equilibrium. Recall the fictitious tavern scene in the movie *A Beautiful Mind* discussed in chapter 2. Ersatz Nash proposed that the best outcome was for everyone to ignore the blonde and go for the brunettes. This strategy, however, does not result in a unique Nash equilibrium. The reason for this is that if everyone goes for a brunette, it will be in someone's best interest to go for the blonde. This game has multiple Nash equilibria, but ignoring the blonde is not one of them. To see why, consider Fig. 4.10.

Assume that there are three students (A, B, and C) and two strategies (*Blonde* and *Brunette*). There are eight possible strategy profiles. The payoffs for {*Blonde*, *Blonde*, *Blonde*} is (0, 0, 0) since they block each other and no one gets the blonde. The

		C			
		Blonde		*Brunette*	
		B		B	
		Blonde	*Brunette*	*Blonde*	*Brunette*
A	*Blonde*	(0, 0, 0)	(0, **1**, 0)	(0, 0, **1**)	**(2, 1, 1)**
	Brunette	(**1**, 0, 0)	**(1, 1, 2)**	**(1, 2, 1)**	(**1**, 1, 1)

Payoffs: (A, B, C)

Fig. 4.10.

payoffs for {*Blonde*, *Brunette*, *Blonde*} is (0, 1, 0) since A and C block each other and B gets a date with a brunette. The payoffs for {*Brunette*, *Brunette*, *Blonde*} is (1, 1, 2) since A and B get dates with a brunette, while C dates the blonde. This game has three possible Nash equilibria: {*Brunette*, *Brunette*, *Blonde*}, {*Brunette*, *Blonde*, *Brunette*}, {*Blonde*, *Brunette*, *Brunette*}. In each, someone dates the blonde.

In games involving multiple Nash equilibria, it may be difficult to predict the strategies of the other players without more information concerning the nature of the relationship between the players, the context within which the game is being played, and whether cooperation between the players is possible, or even desirable. A common understanding of the game's environment may result in what Thomas Schelling (1960) called a *focal-point equilibrium*, which occurs when all players fixate on the same strategy profile. When this is not the case, a player may use a selection criterion other than the highest payoff. One possibility is to adopt a *maximin strategy*.

Maximin (Secure) Strategy

We have so far assumed that a player's strategy choices were based on obtaining the best payoff. A player who does not have a dominant strategy, however, may adopt a **maximin strategy** (also known as a *secure strategy*) if the aim is to avoid the worst possible outcome. It is a strategy that maximizes a player's minimum payoffs. Consider the game depicted in Fig. 4.11 in which neither player has a dominant strategy.

		B		
		B1	*B2*	*B3*
A	*A1*	(1, **5**)	(**3**, 3)	(2, 4)
	A2	(2, 1)	(1, **5**)	(**5**, 4)
	A3	(**3**, 3)	(2, 4)	(4, **5**)

Payoffs: (A, B)

Fig. 4.11.

Suppose that A wants to avoid the worst payoff. This player begins by examining the payoffs from each possible strategy. The worst payoff if A moves *A1* or *A2* is 1. The worst payoff if A moves *A3* is 2. A's maximin strategy is *A3* since it results in the best (maximum) of these worst (minimum) payoffs. Similarly, the worst payoff if B moves *B1* is 1. The worst payoff if B moves *B2* is 3. The worst payoff if B moves *B3* is 4. Player B's maximin strategy is *B3* since it results in the best of the worst payoffs. If both players adopt their maximin strategies, the resulting strategy profile is {*A3*, *B3*} with payoffs of (4, 5).

In some cases, a player may adopt a maximin strategy even though it appears that the other player will adopt a dominant strategy. The game depicted in Fig. 4.12 summarizes the strategies and profits of two competing firms. In this game, B has a dominant *B2* strategy. If A believes B is a profit maximizer, A's best response is A2. The resulting Nash equilibrium strategy profile is {*A2*, *B2*}.

		B	
		B1	*B2*
A	*A1*	**(100,** 0)	(100, **100)**
	A2	(−1,000, 90)	**(200, 100)**

Profits: (A, B)

Fig. 4.12.

Suppose, however, that B's intention is to impose a large loss on A, possibly causing it to file for bankruptcy. To accomplish this, B may signal its intention to move *B2*, but instead moves *B1*. If B's bluff is successful, the payoff to A from the resulting strategy profile {A2, *B1*} is –1,000. This example highlights the importance of correctly interpreting a rival's intentions. In risky games with uncertain payoffs, a player may choose to adopt a secure strategy, even when it does not appear to be in his or her best interest to do so. In this game, A's best response is a maximin *A1* strategy.[2]

There is no general relationship that links secure and Nash equilibrium strategy profiles. In two-player static games in which both players have a strictly-dominant strategy, however, a strategy profile that includes one player's maximin payoff also includes the other player's maximin payoff

Practice Exercises

4.1 Suppose a group of co-workers decide to have lunch at a Japanese restaurant. They have agreed to share equally the lunch bill. Since a small bill is preferred to a large bill, each worker will order one of the least expensive items on the menu. Do you agree? Explain.

4.2 What is the difference between a strictly-dominant strategy and a weakly-dominant strategy?

4.3 Explain the difference between a strictly-dominant strategy and an iterated-strictly-dominant strategy.

4.4 Under what circumstances might a player prefer a maximin strategy to a dominant strategy?

4.5 Consider the two-player, one-time, static game involving players A and B. The four strategy profiles for this game are {*Yes*, *Heads*}, {*Yes*, *Tails*}, {*No*, *Heads*}, and {*No*, *Tails*}. The respective (A, B) payoffs are (5, 10), (6, 1), (–3, 3), and (4, –4).

a. Does either player have a strictly-dominant strategy?
b. Does this game have a Nash equilibrium strategy profile?
c. Is there any coalition that will make both players better off?

4.6 Argon Airlines and Boron Airways compete for passengers travelling the Boston–Albany–Buffalo corridor. Both airlines are considering offering discount air fares during the traditionally slow month of February. The four strategy profiles for this static game are {*Discount*, *Discount*}, {*Discount*, *No discount*}, {*No discount*, *Discount*}, and {*No discount*, *No discount*}. The respective (Argon, Boron) profits in millions of dollars are (2, 3), (7, 1), (1, 6), and (3, 2).

a. If larger payoffs are preferred, does either firm have a strictly-dominant strategy?
b. What is the Nash equilibrium strategy profile for this game? Explain.

4.7 Consider the static game depicted in the accompanying figure.

a. For what values of y is strategy *A2* dominant?
b. For what values of x is strategy *B1* dominant?
c. For what values of y is the strategy profile {*A2*, *B2*} a Nash equilibrium?
d. For what values of x is the strategy profile {*A1*, *B2*} a Nash equilibrium?

		B	
		B1	*B2*
A	*A1*	$(3, 6-x)$	$(5, 4)$
	A2	$(4, 5)$	$(6-y, 3)$

Payoffs: (A, B)

Fig. PE4.7.

4.8 Firm A is considering entering a market in which firm B is a monopolist. B must decide how to respond. If A enters, B can adopt a *predatory* or a *competitive* pricing strategy. If B moves *predatory*, A will lose \$5 million and B will earn \$1 million. If B moves *competitive*, A will earn \$3 million and B will earn \$7 million. Finally, if A's move is to *stay out*, B will earn \$14 million.

a. Does either firm have a dominant strategy?
b. What is the Nash equilibrium strategy profile for this game?

4.9 Tom and Brandy are competitors who sell gourmet tea. Expected profits are summarized in the accompanying figure.

a. Does either player have a strictly-dominant strategy?
b. Does either player have a strictly-dominated strategy?
c. Does this game have a unique Nash equilibrium strategy profile?

		Brandy		
		Pekoe	*Oolong*	*Green*
Tom	*Pekoe*	(150, 150)	(300, 250)	(350, 200)
	Oolong	(300, 100)	(200, 200)	(150, 300)
	Green	(250, 50)	(260, 150)	(300, 300)

Profits: (Tom, Brandy)

Fig. PE4.9.

4.10 Consider the three-player game depicted in the following figure. If larger payoffs are preferred, does this game have a unique Nash equilibrium?

		C								
		C1			*C2*			*C3*		
		B			**B**			**B**		
		B1	*B2*	*B3*	*B1*	*B2*	*B3*	*B1*	*B2*	*B3*
	A1	(1, 9, 3)	(2, 6, 4)	(8, 5, 9)	(6, 6, 5)	(1, 5, 3)	(3, 4, 8)	(9, 2, 6)	(7, 3, 2)	(3, 1, 9)
A	*A2*	(4, 2, 0)	(6, 4, 0)	(8, 2, 3)	(7, 1, 0)	(2, 2, 5)	(4, 0, 1)	(5, 4, 1)	(7, 5, 1)	(4, 4, 2)
	A3	(3, 1, 4)	(3, 6, 7)	(9, 5, 1)	(8, 4, 5)	(1, 6, 6)	(5, 4, 2)	(9, 8, 2)	(8, 7, 5)	(3, 9, 3)

Payoffs: (A, B, C)

Fig. PE4.10.

4.11 Suppose that the Carolina Panthers and the New England Patriots football teams are in the fourth quarter of a game with seconds left on the clock. There is enough time left for one last play. The score is Panthers 12 and the Patriots 17. The Patriots have the ball on the Patriots 8-yard line. New England must score a touchdown for 6 points to win. Each side must decide its final play. The four strategy profiles for this static game are {*Pass defense, Pass offense*}, {*Pass defense, Run offense*}, {*Run defense, Pass offense*}, and {*Run defense, Run offense*}. The respective (Panthers, Patriots) probabilities of winning are (65, 35), (27, 73), (74, 26), and (30, 70).

a. Does either team have a dominant strategy?
b. Does this game have a Nash equilibrium?
c. What is the strategy profile if teams both adopt a maximin strategy?

4.12 Two radio stations, KRZY and KRUD, are considering changing their programming formats to Rock, Country, or Talk. Research indicates that the market shares for these three formats are 40, 30, and 20 percent, respectively. If each station selects the same format, they will split that segment of the market. If the stations choose different formats, they will get the total share of that segment. The stations' advertising revenues are proportional to their market shares, which are summarized in the accompanying figure.

a. If larger market shares are preferred, does either radio station have a dominant strategy?
b. Does this game have a Nash equilibrium?
c. What is the strategy profile if both radio stations adopt a maximin strategy?

		KRUD		
		Rock	*Country*	*Talk*
	Rock	(20, 20)	(40, 30)	(40, 20)
KRZY	*Country*	(30, 40)	(15, 15)	(30, 20)
	Talk	(20, 40)	(20, 30)	(10, 10)

Market shares: (KRZY, KRUD)

Fig. PE4.12.

4.13 Consider the competitive, one-time, static game involving players A and B. There are four strategy profiles: {*A1*, *B1*}, {*A1*, *B2*}, {*A2*, *B1*}, and {*A2*, *B2*}. The respective payoffs are (100, 100), (75, 75), (–100, 90), and (200, 200).

a. If larger payoffs are preferred, does either player in this game have a dominant strategy?
b. What is the strategy profile for this game if both players adopt a secure strategy?
c. What is the most likely outcome if neither player adopts a secure strategy?

Notes

1. Formally, suppose a player's payoff function u_i is defined over the set of strategies S, where $u_i(s)$ is player *i*'s payoff given strategy $s_i \in S_i$. For every two strategy profiles *s*, where $s' \in S$ in a one-time, two-player, static competition game, $u_1(s) > u_1(s')$ if, and only if, $u_2(s) < u_2(s')$.

2. Formally, a strategy $s_i \in S_i$ is a secure strategy if player *i* is guaranteed a payoff of *at least* $w_i(s_i)$, that is, s_i satisfies $\max_{si} \in {}_{Si}\, w_i(s_i)$. This *maximin* payoff may be written

$$\max_{s_i \in S_i} \min_{s_j \in S_j} u_i(s_i, s_j).$$

CHAPTER FIVE

Coordination Games

Introduction

We also saw in our discussion of coalition games that having a dominant strategy does not guarantee that a player will receive the best payoff. When this happens, it may be in the players' best interest to cooperate, although each player has an incentive to violate the agreement. We saw that when a player in a one-time, pure-strategy, static competition game has a dominant strategy, a unique Nash equilibrium strategy profile will emerge. If neither player has a dominant strategy, the game may have multiple Nash equilibria, or may have none at all. When this happens, a player might adopt a maximin strategy.

In this chapter we will examine games in which players coordinate their strategies to maximize their joint payoffs. Players in a **coordination game** derive utility from the benefits received by other players. It may also be possible for players to combine and redistribute payoffs. Unlike coalition games, the players in coordination games do not have an incentive to defect. Coordinated strategies result in win-win outcomes.

Battle-of-the-Sexes Game

A well-known example of the benefits of coordinated strategies is the battle-of-the-sexes game. In this game, Rhett and Scarlett are trying to decide whether to spend the evening together or apart. Scarlett prefers to have dinner at an expensive restaurant and Rhett would like to eat barbecue at a neighborhood tavern. Both, however, prefer to spend the evening together rather than apart. This one-time, static game is depicted in Fig. 5.1. The payoffs are utility indices that capture the relative intensity of each player's preferences.

		Rhett	
		Tavern	*Restaurant*
Scarlett	*Tavern*	**(10, 20)**	(6, 10)
	Restaurant	(8, 6)	**(20, 8)**

Utility: (Scarlett, Rhett)

Fig. 5.1.

This battle-of-the-sexes game has two Nash equilibrium strategy profiles: {*Tavern*, *Tavern*} and {*Restaurant*, *Restaurant*}. Since Rhett and Scarlett are concerned with each other's happiness, the best outcome results when they coordinate their plans. In coordination games, the players should not be viewed as rivals, but partners whose objective is to maximize their joint payoffs, which means that they will have dinner at the neighborhood tavern. If this game is played repeatedly, Rhett and Scarlett may agree to have dinner at a tavern this week and dinner at a fancy restaurant the next. Either way, defection by either player results in a lose-lose outcome. Win-win situations in coordination games are frequently encountered in business, such as when collective bargaining results in profit sharing plans or when firms enter into strategic alliances.

Solved Exercise

Rival firms A and B are considering whether to charge a high price *or a* low price *for their products. The four strategy profiles are {High price, High price}, {High price, Low price}, {Low price, High price}, and {Low price, Low price}. The respective (A, B) payoffs in millions of dollars are (6, 8), (8, 7), (5, 7), and (7, 13).*

a. *If larger payoffs are preferred, what is the Nash equilibrium strategy profile?*
b. *Is there any coalition that will make both firms better off?*
c. *Suppose the firms agree to pool and share their profits. Is coordinating strategies stable if neither firm has a bargaining advantage?*
d. *Is coordination possible if A can extract 70 percent of the joint profits?*

Solution

a. *Since* A *has a strictly-dominant* high price *strategy, the Nash equilibrium strategy profile is {High price, High price}.*
b. *The strategy profile {Low price, Low price} results in a better payoff for both firms. This coalition is unstable since both firms have an incentive to defect.*
c. *In a coordination game with no dominant players, the most likely strategy profile is {Low price, Low price} since the payoff to each player is $9 million, which is greater for both players than the Nash equilibrium payoffs.*
d. *Coordination is not possible since the payoff to* B *is $6 million, which is less than the Nash equilibrium payoff of $8 million.*

Focal-Point Equilibrium

We have examined situations involving multiple Nash equilibria in which the players explicitly agree to coordinate their moves. In some situations, however, a single strategy may emerge without an explicit agreement due to a shared understanding of the environment within which the game is played. Thomas Schelling referred to this as a **focal-point equilibrium.**[1]

A focal-point equilibrium is sometimes easily explained by examining the payoffs. In other cases, the existence of a focal-point equilibrium is easy to identify, but not easily explained. In 1964, Supreme Court Justice Potter Steward tried to described "hard-core" pornography by observing: "I shall not today attempt further to define the kind of material I understand to be embraced . . . [b]ut I know it when I see it. . . ."[2] A focal-point equilibrium is something like that. Also called a *Schelling-point equilibrium*, this outcome is said to be self-fulfilling because the players believe that this strategy profile is in their self interest.

To illustrate the idea underlying the existence of a focal-point equilibrium, suppose that a father and son become separated at an amusement park. How likely is it that both will think to look for each other at the same location in the event they become separated? This coordination game has multiple Nash equilibrium strategy profiles, such as the main gate or the office of the park police. Will a shared belief system lead father and son to search for each other at the same location?

Avinash Dixit and Barry Nalebuff (1991, ch. 3) pose the following question. Will driving on the left or the right side of the road will result in the best outcome? Specifying the order of moves, strategies, and payoffs is insufficient for determining a best response. The correct strategy depends on where the question is asked. In Beijing, Montreal, New York City, or Paris, the dominant strategy is to drive on the right. In Hong Kong, London, Sydney, or Tokyo, the dominant strategy is to drive on the left. Without knowing location, searching for dominant strategies, or eliminating dominated strategies, will not result in a Nash equilibrium.

Thomas Schelling (1960) illustrated the concept of a focal-point equilibrium by conducting several experiments involving hypothetical situations. In one experiment, Shelling asked pairs of students from New Haven, Connecticut, to separately name a time and place to meet on a given day in New York City. Neither student had ever been placed in this situation before, nor were they permitted to communicate with each other. Both win a prize if choose the same location. If they each guess correctly, both win a prize, otherwise they win nothing. A majority of the respondents proposed meeting at the information booth in Grand Central Terminal, and nearly all of them agreed to meet at 12:00 noon.

Schelling found that the test subjects in his experiments tended to choose focal points, even when it was not in their individual best interest to do so. For example, in another experiment, players were identified as A or B. On the flip of a coin, the players were simultaneously asked to called "heads" or "tails." If both players called "heads," A received $3 and B received $2. If both called "tails," A received $2 and B received $3. If one player called "heads" and the other called "tails," both received nothing. Schelling found that 73 percent of player respondents chose "heads" when given the role of A. Surprisingly, 68 percent of players in the role of B still chose "heads" despite the lower payoff.

David Kreps (1987, 1990) conducted experiments involving pairs of college students who were provided with a list of eleven cities (Atlanta, Boston, Chicago, Dallas, Denver, Kansas City, Los Angeles, New York, Philadelphia, Phoenix, and San Francisco). Each city was ranked according to its importance in commerce, science, the arts, and so on. New York, for example, was assigned the highest ranking (100) and Kansas City the lowest (1). The students were asked to assign each city to one of two subsets. Communication between the students was prohibited. The only requirement was that one subset must include Boston and the other subset must include San Francisco. If a city appeared on one list, but not the other, the student listing the city received a dollar amount equal to the city's ranking. If a city appeared on both lists, each lost an amount double the city's ranking. Finally, if all eleven cities were partitioned without duplication, winnings were tripled.

This coordination game had 512 Nash equilibria. In nearly every trial the Boston list included New York and Philadelphia, and, less frequently, Chicago. The San Francisco list invariably included Los Angeles, Phoenix, and Denver, and less frequently Dallas and Kansas City. These assignments appear to have been based on geography, with the Mississippi River being the sorting criterion. In some cases, the geographic sorting criterion was Snowbelt versus Sunbelt states. Sorting may also have been based on cultural norms, regional attitudes and perceptions, history, or whether a city's baseball team was in the American or National League.

Solved Exercise

Consider a one-time static game involving players A and B. This game has four strategy profiles: {A1, B1}, {A1, B2}, {A2, B1}, and {A2, B2}. The respective (A, B) payoffs are (2, 2), (0, 1), (1, 0), and (1, 1).

a. *Does either firm have a dominant strategy?*
b. *Does this game have a unique Nash equilibrium?*
c. *What is the most likely outcome for this game?*

Solution

a. *Neither firm has a dominant strategy. The best response by each player changes with the strategy of the other player.*
b. *This game has two pure-strategy Nash equilibrium strategy profiles: {A1, B1} and {A2, B2}.*
c. *The strategy profile {A1, B1} is a focal-point equilibrium since this results in the best payoff for both players.*

Developing a Theory of Focal-Point Equilibria

There is a substantial body of literature that attempts to identify the conditions whereby a single strategy profile emerges as the most likely in games involving multiple Nash equilibria. These conditions typically have something to do with the context within which the game is played, the nature of the relationships among players, or some shared interest that leads players to "zero in" on a particular strategy profile. Formalizing the conditions for the existence of such a focal-point equilibrium, however, has proven to be elusive. Indeed, game theorists have not even been able to agree on a formal definition of a "focal point."

There appear to be a number of qualitative factors that help to explain its existence. Distribution of payoffs, symmetric notions of equity, efficiency, and fairness appear to be important identifying principles. Moreover, when one identifying principle suggests a unique equilibrium, and another suggests multiple equilibria, the first principle tends to dominate. Alvin E. Roth, J. Keith Murnighan, and Francoise Schoumaker (1988) discovered that implicit agreements may also emerge from a process of adaptive expectations. Observing how players have played the game in the past affects how players play the game in the future.

Framing

In games involving multiple Nash equilibria, notions of equity and fairness appear to be important guiding principles when identifying focal points. Somewhat surprisingly, concepts of fairness and equity often appear to violate the assumption of rational behavior. Studies have shown that the context within which choices are presented is important. **Framing** refers to the idea that people often make decisions based upon a familiar frame of reference. This frame of reference makes it possible for individuals to make comparisons and select from among different alternatives. One such frame of reference is the conventional notion of "fairness."

The idea of framing can be illustrated with the following simple experiment, which is referred to as the *ultimatum game*. Two players are identified as a "giver" or a "taker." The giver is given $10 and is instructed to offer the taker a share. If the taker accepts the offer, the giver and taker keep the proposed distribution. If the taker rejects, both sides receive nothing.

The rationality assumption suggests that the giver should offer the smallest possible share, and the taker should accept since getting something is better than getting nothing. For example, if offers are in dollar increments, the giver should offer $1 and the taker should accept. Surprisingly, in repeated experimental trials givers tended to offer $4 or $5, but never more than $5. Offers of $1, $2, or $3 were nearly always rejected, and offers of $4 were sometimes rejected. Although rejection of these offers appears irrational, the players shared frame of reference seems to suggest a "fair" distribution was preferred.

Practice Exercises

5.1 Explain the difference between a coalition game and a coordination game.

5.2 How does a Nash equilibrium in a coordination game differ from a focal-point equilibrium?

5.3 When can focal points be used to identify a Nash equilibrium?

5.4 If neither player in a one-time, static game has a dominant strategy, multiple Nash equilibria are not possible, but there is always a focal-point equilibrium. Do you agree?

5.5 What should Phoebe and Chloe do if their phone call gets cut off? Should Phoebe call Chloe, or should Chloe call Phoebe? Explain your reasoning.

5.6 Two motorists approach an intersection at the same time. On each corner is a stop sign. Driver A is driving north and driver B is driving west. Which driver should enter the intersection first, or should they enter the intersection at the same time? Does it matter whether a driver goes straight, turns right, or turns left? Does this game have a focal-point equilibrium? Explain.

5.7 Brad and Angelina are trying to decide how to spend an evening. Although Brad prefers the opera and Angelina prefers professional wrestling, they would rather spend the evening together than apart. The four strategy profiles are {*Opera*, *Opera*}, {*Opera*, *Wrestling*}, {*Wrestling*, *Opera*}, and {*Wrestling*, *Wrestling*}. The respective (Brad, Angelina) payoffs are (5, 8), (−30, –20), (–2, –5), and (6, 9).

a. If larger payoffs are preferred, does this game have Nash equilibrium?
b. What is the most likely outcome if Brad and Angelina coordinate their plans?

5.8 Two hunters are stalking hares or stags. Each hunter could stalk stag alone, but because of the size and weight they would only be successful bringing down a stag by coordinating their efforts. Sharing a stag is preferred to sharing hares. The hunters' strategy profiles are: {*Stag*, *Stag*}, {*Stag*, *Hare*}, {*Hare*, *Stag*}, and {*Hare*, *Hare*}. The respective utility payoffs are (6, 5), (0, 1), (2, 0), and (3, 2). Greater utility is preferred.

a. Does either hunter have a dominant strategy?
b. Does this game have a unique Nash equilibrium?
c. Does this game have a focal-point equilibrium?

5.9 Consider the one-time, static game involving players A and B. There are four strategy profiles: {*A1*, *B1*}, {*A1*, *B2*}, {*A2*, *B1*}, and {*A2*, *B2*}. The respective (A, B) payoffs are (74, 84), (98, 72), (60, 50), and (80, 62). Larger payoffs are preferred.

a. What is the Nash equilibrium strategy profile if A and B are competitors?
b. Suppose that A and B coordinate their strategies. What is the most likely strategy profile if A and B have equal bargaining power and neither player can commit to a strategy first?
c. What is the most likely strategy profile if A has twice the bargaining power as B? Explain your answer.

5.10 Consider the one-time, static game depicted in the accompanying figure.

a. What is the Nash equilibrium if the players are competitors?

b. Will the players coordinate their strategies if they have equal bargaining power? What is the most likely strategy profile?

c. Will the players coordinate their strategies if C has twice as much bargaining power as A, who has just as much bargaining power as B? What is the most likely strategy profile? Explain your answer.

		C								
		C1			*C2*			*C3*		
		B			**B**			**B**		
		B1	*B2*	*B3*	*B1*	*B2*	*B3*	*B1*	*B2*	*B3*
	A1	(1, 9, 3)	(2, 6, 4)	(8, 5, 9)	(6, 6, 5)	(1, 5, 3)	(3, 4, 8)	(8, 2, 6)	(7, 3, 2)	(3, 1, 8)
A	*A2*	(4, 2, 0)	(6, 5, 6)	(8, 2, 3)	(7, 1, 0)	(2, 2, 1)	(4, 0, 1)	(5, 4, 1)	(7, 5, 1)	(4, 4, 2)
	A3	(3, 1, 4)	(3, 6, 7)	(9, 5, 1)	(8, 4, 5)	(1, 6, 6)	(5, 4, 2)	(9, 8, 3)	(8, 8, 3)	(3, 9, 3)

Payoffs: (A, B, C)

Fig. PE5.10.

Notes

1. In 2005, political scientist Thomas Schelling shared the Nobel Prize in economics with mathematician Robert Aumann for their work in game theory. The Nobel committee lauded their work for "having enhanced our understanding of conflict and cooperation." Schelling is probably best known for his innovation of the payoff matrix, which is used for analyzing static games. Schelling once quipped: "If I am ever asked whether I ever made a contribution to game theory, I shall answer yes. Asked what it was, I shall say the invention of staggered payoffs in a matrix . . . I did not suppose that the invention was patentable, so I made it freely available and hardly anybody except my students takes advantage. I offer it to you free of charge."

2. Jacobellis v. Ohio, 378 U.S. 184.197 (164).

CHAPTER SIX

Infinitely-Repeated Games

Introduction

Proctor & Gamble (P&G) and Unilever are giants in the household-products industry. Cincinnati-based P&G brands include Actonel, Always/Whisper, Bounty, Charmin, Crest, Downy, Folgers, Iams, Olay, Pampers, Pantene, Pringles, Tide, and Wella. Unilever is jointly owned by Netherlands-based Unilever N.V. and United Kingdom-based Unilever PLC. The company's brand names include Calvin Klein, Birds Eye, Dove, and Lipton. For decades, P&G and Unilever were locked in an intense global competition. In recent years, however, these companies appear to have come to an understanding not to slash prices, introduce new product lines without prior notice, or launch advertising campaigns involving product comparisons.

We learned in our discussion of one-time, static coalition games that it is frequently in the best interest of players to cooperate, but each has an incentive to defect resulting in a less-than-optimal outcome. So what accounts for the seeming longevity and stability of the P&G-Unilever relationship?

Coalitions

A coalition involves two or more players who cooperate to achieve a mutually beneficial outcome. The most recognizable manifestation of business **collusion** is the **cartel**, which is an explicit formal agreement by firms to coordinate their activities to increase industry profits. One of the best-known examples of a cartel is the Organization of Petroleum Exporting Countries (OPEC), which attempts to influence world oil prices by regulating output. Unlike the P&G/Unilever relationship, OPEC agreements often break down because smaller members routinely exceed their production quotas to bolster oil export earnings.

Collusion need not be a formal overt agreement, but may evolve from the players' shared understanding of the strategic environment. Secret agreements are frequently meant to hide illegal activities. In the 1950s, for example, General Electric and Westinghouse rigged seal-bid auctions to sell turbine generators to several government-owned utilities (Fuller, 1962; Porter, 1980). After the conspiracy was uncovered and dismantled, the prices of large turbines fell by fifty percent between 1958 and 1963.

Although business collusion typically aims to raise prices by restricting supply, it also involves many other business activities, such as reducing service quality (which until recently was practiced by European airlines), limiting advertising expenditures, and restricting firms' sales territory, such as occurred in the European chemical industry in the 1920s. The agreement gave ICI (Imperial Chemical Industries, PLC) the exclusive distribution rights in Great Britain and the Commonwealth countries, DuPont in the United States, and German firms on the European continent. This practice was declared illegal in the early 1930s.

To protect consumers against such anti-competitive business practices, the U.S. government enacted a host of antitrust legislation, including the Sherman Act (1890), the Clayton Act (1914), and the Federal Trade Commission Act (1914). In Europe, collusion was outlawed by Article 85 of the Treaty of Rome.

In the absence of an effective enforcement mechanism, defection is a distinguishing feature of coalition games that are played just once. In business, however, ongoing competition between and among rival firms is the rule rather than the exception. Is it possible for a coalition that is unstable in a one-time game to constitute a Nash equilibrium if the game is played repeatedly?

Repeated Static Games

One possible explanation for coalitional stability may be found in the dynamics of ongoing strategic relationships. Competition in business is more appropriately analyzed as a static game that is played repeatedly. In the remainder of this chapter, we will analyze **infinitely-repeated static games**, which are static games that played over and over without end. The purpose is to develop a framework for analyzing more realistic **finitely-repeated static games** with certain and uncertain ends, which will be discussed in the next chapter. As before, we will continue to assume that all players have complete information.

What is it about the ongoing relationship between P&G and Unilever that makes cooperation mutually beneficial? An insight into the dynamics of repeated games may be gleaned by considering the static game depicted in Fig. 6.1. This one-time, coalition game has two possible pure-strategy Nash equilibrium strategy profiles, although {*A2*, *B2*} is a focal-point equilibrium. It should be readily apparent, however, that both players would be better off by adopting the coalition strategy profile {*A1*, *B1*}. In a one-time game this collation will collapse since both players have an incentive to defect. Is it possible, however, for the strategy profile {*A1*, *B1*} to constitute a Nash equilibrium if this game is played repeatedly?

		B1	B2	B3
	A1	(20, 20)	(4, **25**)	(0, 0)
A	A2	**(25**, 4)	**(8, 8)**	(0, 0)
	A3	(0, 0)	(0, 0)	**(1, 1)**

(Column player: **B**; row player: **A**)

Payoffs: (A, B)

Fig. 6.1.

There is a fundamental difference between coalition games that are played once and those that are played repeatedly. In a one-time, coalition game retaliation in the event of defection is an empty threat because there is no further game play. This is clearly not the case when games are played repeatedly. The threat of retaliation in the event of defection fundamentally changes the nature of strategic coalitions.

Trigger Strategies

Suppose that the coalition game depicted in Fig. 6.1 is infinitely repeated and A and B form a coalition and agree to the strategy profile {*A1*, *B1*}. Suppose that A makes known to B the following strategy for all future game play:

1. Move *A1* in stage 1;
2. Continue to move *A1* if B moves *B1*;
3. If B does not move *B1*, move *A2* in the next stage and all subsequent stages.

A's strategy of moving *A2* should B defect is called a **trigger strategy**, which is a move by one player in response to an unanticipated move by another player. Defection by B "triggers" a change in strategy by A. If, in fact, it is part of A's strategy to never cooperate again, then *A2* may also be described as a **grim strategy**. Will A's trigger strategy discourage defection by B?

As it turns out, A's trigger strategy to prevent defection is probably unnecessary since it appears to be in B's best interest to cooperate. To see why, suppose that A moves *A1* in stage 1 and B immediately defects. B's payoff in stage 1 is 25 and A's payoff is 4. A "punishes" B by switching to *A2*. The strategy profile in stage 1 is {*A1*, *B2*} and {*A2*, *B2*} in all subsequent stages. The undiscounted payoff to B from defection is 25 + 8 + 8 + . . . compared with 20 + 20 + 20 + . . . by cooperating.[1] It would appear that defection is a *strictly-dominated strategy*. Because of the symmetry of the game, cooperation appears to constitute a dominant strategy for A as well.

What is interesting about the coalition game depicted in Fig. 6.1 is that {*A1*, *B1*} does not appear to constitute a Nash equilibrium if the game is played once, but may be a Nash equilibrium strategy profile if the game is infinitely repeated. Unfortunately, appearances can be deceiving because the payoffs from alternative strategy profiles have not been properly evaluated.

Evaluating Payoffs in Infinitely-Repeated Games

To determine if it is in a player's best interest to defect, we need to compare the present value of the stream of payoffs from defection to the present value of the stream of payoffs from cooperation.[2] The present value of an infinite stream of future payoffs for player *j* is

$$PV^j = \frac{\pi_0^j}{(1+i)^0} + \frac{\pi_1^j}{(1+i)^1} + \frac{\pi_2^j}{(1+i)^2} + \cdots, \tag{6.1}$$

where π_t is the payoff for player *j* in each stage of the game and *i* is the interest rate used to discount future payoffs. If the payoffs in each stage of the game are equal, Eq. (6.1) becomes

$$PV^j = \pi^j \sum_{t=0}^{\infty} \left(\frac{1}{1+i}\right)^t, \tag{6.2}$$

where the term in the parentheses is the **discount rate**. By definition, the discount rate and the interest rate vary inversely. Eq. (6.2) may be rewritten as[3]

$$PV^j = \pi^j \left(\frac{1+i}{i}\right). \tag{6.3}$$

The present value of the payoffs to player *j* from a coalition is

$$PV_C^j = \pi_C^j \left(\frac{1+i}{i}\right), \tag{6.4}$$

where $\pi_C^{\ j}$ is the payoff to player *j* from cooperating. It may be demonstrated that the present value of the payoffs to a player from unilateral defection is equal to the *net* one-time gain ($\pi_V^{\ j} - \pi_N^{\ j}$) plus the present value of the Nash equilibrium payoffs ($PV_N^{\ j}$), that is

$$PV_D^j = \left(\pi_D^j - \pi_N^j\right) + PV_N^j = \left(\pi_D^j - \pi_N^j\right) + \pi_N^j \left(\frac{1+i}{i}\right). \tag{6.5}$$

It will pay for a player to defect whenever the present value of the stream of payoffs from defection is greater than the present value of the stream of payoffs from cooperation, that is

$$(\pi_D^j - \pi_N^j) + PV_N^j > PV_C^j. \tag{6.6}$$

Expanding Inequality (6.6) we obtain

$$(\pi_D^j - \pi_N^j) + \pi_N^j\left(\frac{1+i}{i}\right) > \pi_C^j\left(\frac{1+i}{i}\right). \tag{6.7}$$

The results may be simplified to derive a defection rule for a two-player, infinitely-repeated, static game. Rearranging Inequality (6.7) we obtain

$$\frac{\pi_C^j - \pi_N^j}{\pi_D^j - \pi_C^j} < i. \tag{6.8}$$

If Inequality (6.8) is satisfied, the players will have a unilateral incentive to defect. If the left side of the inequality is greater than the interest rate, the players will have an incentive to form a coalition. If the left side of the inequality is equal to the interest rate, players will be indifferent between cooperation and defection.

Solved Exercise

Consider the game in Fig. 6.1. If the interest rate is 5 percent, calculate the present value of the stream of future payoffs from:

a. *cooperation;*
b. *unilateral defection;*
c. *the sucker's payoff.*
d. *Based on the above answers, is the coalition strategy profile {A1, B1} stable?*
e. *For what interest rates can we expect unilateral defection by either player?*

Solution

a. *The present value of the stream of payoffs from cooperation is*

$$PV_C^j = 20\left(\frac{1.05}{0.05}\right) = 420. \tag{6.9}$$

b. *The present value of the stream of payoffs from defection is*

$$PV_D^j = 25 + \frac{8}{(1.05)^1} + \frac{8}{(1.05)^2} \cdots = (25 - 8) + 8\frac{(1.05)}{0.05} = 185. \tag{6.10}$$

c. *The present value of the sucker's payoff is*

$$PV_S^A = (4-8) + 8\frac{(1.05)}{0.05} = 164. \quad (6.11)$$

$PV_S^{\,A}$ is the present value of the payoffs to A when B defects. The value 4 – 8 = –4 is the loss to A by being faithful compared to the payoffs by not cooperating. This is the sucker's payoff in the first stage.

d. *Since $PV_C^{\,j} > PV_D^{\,j}$, the coalition strategy profile {A1, B1} is stable.*

e. *From Inequality (6.11),*

$$\frac{\pi_C^j - \pi_N^j}{\pi_V^j - \pi_C^j} = \frac{20-8}{25-20} = 2.40. \quad (6.12)$$

Each firm will have a unilateral incentive to defect for any interest rate greater than 240 percent.

Practice Exercises

6.1 Explain the prisoner's dilemma in a one-time, static game.

6.2 When will the prisoner's dilemma in a one-time game be overcome in an infinitely-repeated, static game?

6.3 When is a credible threat similar to a trigger strategy? How are they different?

6.4 When will a trigger strategy deter defection in a coalition game?

6.5 Trigger strategies always prevent defection in infinitely-repeated, static coalition games. Do you agree? Explain.

6.6 What is a grim strategy?

6.7 When will coalitions be stable in infinitely-repeated games?

6.8 Tacit collusion is more stable than explicit collusion. Do you agree? Why?

6.9 Consider again the static game depicted in Fig. 6.1. Suppose that A adopts the following strategy:

1. Move *A1* in stage 1;
2. If B moves *B1* in stage 1, move *A1* in stage 2;
3. If B does not move *B1* in stage 1, move *A3* in stage 3;
4. Continue to move A3 unless B agrees to move *B1* in stage 4, and so on.
    a. What are the payoffs if B moves *B1* in stage 1, defects in stage 2?
    b. What are the payoffs if B moves *B1* in stage 1, defects in stage 2, reforms the coalition in stage 3, defects in stage 4, and so on?
    c. Will A's trigger strategy keep B from defecting?

6.10 Orange and Blue companies are considering *aggressive* and *reactive* investment strategies. The four strategy profiles are {*Aggressive*, *Aggressive*}, {*Aggressive*, *Reactive*}, {*Reactive*, *Aggressive*}, and {*Reactive*, *Reactive*}. The respective payoffs in thousands of dollars are (500, 500), (100, 600), (600, 100), and (150, 150). Suppose that the interest rate in this infinitely-repeated game is 7 percent.

a. What is the present value of the Nash equilibrium payoffs?
b. What is the present value of the payoffs by forming a coalition?
c. What is the present value of the payoffs from unilateral defection?
d. Based on your answers to the above, is a coalition stable?
e. For what interest rates will the coalition strategy profile break down?

6.11 Consider the static game in Fig. 6.1. Suppose that the interest rate is 300 percent.

a. What is the present value of future payoffs if the firms do not form a coalition?
b. What is the present value of future payoffs if the firms form a coalition?
c. What is the present value of future payoffs from unilateral defection?
d. Based on your answers to the above, is a coalition stable?

6.12 Tsunami Corporation and Cyclone Company are considering a change in their pricing policies. There are four strategy profiles: {*High price*, *High Price*}, {*High price*, *Low price*}, {*Low price*, *High price*}, and {*Low price*, *Low price*}. The corresponding payoffs in millions of dollars are (6, 6), (4, 8), (8, 4) and (5.5, 5.5). Larger payoffs are preferred.

a. What is the Nash equilibrium strategy profile if this game is played one time?
b. Suppose that the interest rate is seven percent and the game is infinitely repeated. What is the present value of future payoffs if the players form a coalition?
c. What is the present value of future payoffs if the companies do not cooperate?
d. What is the present value of future profits from unilateral defection?
e. Based on the above answers, is a coalition stable?
f. For what interest rates will the coalition strategy profile break down?

6.13 Suppose the game in Fig. 2.3 is infinitely repeated and the interest rate is 5 percent.

a. What is the present value of future profits if the firms do not cooperate?
b. What is the present value of future profits to each firm from cooperation?
c. What is the present value of future profits from defection?
d. Do these firms have an incentive to cooperate or unilaterally defect?
e. For what interest rates will the coalition strategy profile break down?

6.14 Suppose that profits earned in the previous exercise are reinvested at an annual interest rate of 25 percent.

a. What is the present value of future payoffs by not cooperating?
b. What is the present value of future payoffs by cooperating?
c. What is the present value of future payoffs from unilateral defection?
d. Is a coalition possible based on the above present values?

Notes

1. Strictly speaking, this is only true for a very large (not infinite) number of rounds (why?).

2. Discounting is necessary because of the time value of money. That is, \$1 received or expended today is worth more than \$1 received or expended tomorrow because the \$1 may be invested and earn a rate of return of *i*.

3. See Appendix A for a derivation of this result.

CHAPTER SEVEN

Finitely-Repeated Games

Introduction

We have thus far considered the two extreme versions of static coalition games with complete information: those that are played just once, and those that are infinitely repeated. We learned that an unstable coalition in a one-time, static game is a Nash equilibrium in an infinitely-repeated game when the present value of future payoffs from a coalition is greater than the present value of future payoffs from unilateral defection. In reality, however, there is no such thing as an infinitely-repeated game.

Finitely-Repeated Games with a Certain End

In this chapter, we will examine *finitely-repeated, static games* in pure strategies. There are two types: those in which the end of the game is known with certainty, and those that are not. We will begin our discussion by examining finitely-repeated, static games with a certain end. Consider the static coalition game in Fig. 6.1. Recall that this one-time game has two Nash equilibrium strategy profiles, although the most likely outcome is {*A2*, *B2*}. Both players are better off, however, by adopting the strategy profile {*A1*, *B1*}. This coalition is unstable, however, since both players have a unilateral incentive to defect.[1]

We saw in the last chapter that the strategy profile {*A1*, *B1*} in Fig. 6.1 may, in fact, constitute a Nash equilibrium in an infinitely-repeated static game if the present value of future payoffs from this coalition is greater than the present value of future payoffs from unilateral defection. It depends on the interest rate. From Inequality (6.8), the players have a unilateral incentive to defect whenever the interest rate is greater than 240 percent. But, what if this game is played a finite number of times? Is it still in the best interest of both players to form an {*A1*, *B1*} coalition?

Suppose that the game in Fig. 6.1 is played two times. The strategy profile {*A2*, *B2*} is a Nash equilibrium in both stages of this game. The reason for this is that the undiscounted payoff to each player is 8 + 8 = 16. The undiscounted payoff to each player from the strategy profile {*A1*, *B1*} is 20 + 20 = 40 > 16. Although the players have an incentive to move {*A1*, *B1*} in both stages, this strategy profile cannot be part of a Nash equilibrium. This is because both players have an incentive to defect in stage 2 since 20 + 25 > 20 + 20. But, if both players defect, the undiscounted payoff is 20 + 8 = 28 < 20 + 20. We are left to determine whether {*A1*, *B1*} in the first stage and {*A2*, *B2*} in the second stage is a Nash equilibrium.

The strategy profile {*A1*, *B1*} in stage 1 and {*A2*, *B2*} in stage 2 is also not a Nash equilibrium. Player B has a unilateral incentive to defect in stage 1 because the undiscounted payoff is 25 + 8 > 20 + 8. Because of the symmetry of the payoffs, A has the same unilateral incentive to defect. Thus, the only strategy profile that constitutes a Nash equilibrium is {*A2*, *B2*} in each stage.

Is there a trigger that will deter defection? Suppose that A adopts the following trigger strategy:

1. Move *A1* in stage 1;
2. If B moves *B1* in stage 1, move *A2* in stage 2;
3. If B does not move *B1* in stage 1, move *A3* in stage 2.

The rationale behind this trigger strategy is that if A moves *A1* in stage 1 and B defects, A will punish B by moving *A3* in stage 2. If B moves *B2* in stage 2 after defecting in stage 1, the undiscounted payoff is 25 + 0 < 20 + 8. It would appear to be in B's best interest not to defect. Since the payoffs are symmetrical, it is in B's best interest to adopt the same trigger strategy.

Does this mean that {*A1*, *B1*} in stage 1 and {*A2*, *B2*} in stage 2 is a Nash equilibrium if both the players adopt a grim {*A3*, *B3*} strategy? No, because {*A3*, *B3*} is an empty threat. Recall that if B unilaterally defects, the resulting strategy profile is {*A1*, *B2*}. If B moves *B2* in stage 2, the undiscounted payoff to A by moving *A3* is 4 + 0, which is less than 4 + 8 = 12 by moving *A2*. Thus, moving *A3* in stage 2 is not credible. The only strategy profile that constitutes a Nash equilibrium is {*A2*, *B2*} in both stages.

End-of-Game Problem

The use of trigger strategies in infinitely-repeated, static games can introduce an element of stability into coalitions. Trigger strategies in finitely-repeated static coalition games with a certain end are ineffective, however, because the last stage is a one-time game. What is more, how a player moves in the last stage determines how that player will move in the first stage, and all subsequent stages. The **end-of-game problem** transforms a finitely-repeated static game with a certain end into a series of one-time, static games.

An interesting real-world example of the end-of-game problem involves a worker who gives notice at the close of business Thursday of her intention to leave at the close of business on Friday. As a rule, people work hard when the marginal benefit of diligence is greater than the marginal benefit of negligence. Given the end-of-game problem, will a worker be diligent on Friday when there is no possibility of retribution by management thereafter?

How will the worker's job performance be affected if she gives notice at the close of business on Wednesday? Friday is now the final stage of a two-stage static game. Thursday and Friday are two one-time games, which suggest that the worker will be negligent on Thursday as well. Indeed, if the worker gives two-week notice, each day becomes a one-time static game in which the worker's dominant strategy is negligence.

What can management do to avoid the end-of-game problem? Management could fire the worker on the spot. This could be counterproductive, however, since the worker might change her strategy from giving two-week notice to quitting at the close of business on Friday. This could make it difficult to find a replacement at short notice, which could disrupt the firm's operations. Moreover, dismissing an employee without cause could lead to morale and legal problems. An alternative approach would be to extend the manager-worker relationship beyond the quit date, such as helping the worker find a new job by providing letters of recommendation. It will be in the worker's best interest to avoid burning bridges by goofing off during her final days on the job.

Finitely-Repeated Game with an Uncertain End

A finitely-repeated game with a certain end collapses into a series of one-time, static games. If the end of the game is uncertain, however, there is no end-of-game problem since the players know the game will end, but not when. Because of this, a trigger strategy that threatens retaliation may deter defection. The analysis is similar to that for infinitely-repeated, static games discussed in the previous chapter.

Although the end of the game is unknown, we assume that the players are able to assign a subjective probability to the next stage being the last. Suppose that the probability that the game will end in the next stage is $0 \leq p \leq 1$. Thus, $(1 - p)$ is the probability that the game will be played in stage 2; $(1 - p)(1 - p) = (1 - p)^2$ that it will be played in stage 3, and so on. In general, the probability that the game will be played in stage t is $(1 - p)^{t-1}$. Thus, the undiscounted expected value of future payoffs for the jth player is

$$E(\pi) = \pi_1^j + \pi_2^j(1-p) + \pi_3^j(1-p)^2 + \pi_4^j(1-p)^3 + \cdots. \tag{7.1}$$

From Eq. (7.1), the present value of future payoffs becomes

$$PV^j = \pi_1^j\left(\frac{1-p}{1+i}\right)^0 + \pi_2^j\left(\frac{1-p}{1+i}\right)^1 + \pi_3^j\left(\frac{1-p}{1+i}\right)^2 + \cdots. \tag{7.2}$$

If the payoff is the same in each period, the present value of an infinite stream of future payoffs when the players do not form a coalition is

$$PV_N^j = \pi_N^j\left(\frac{1+i}{i+p}\right). \tag{7.3}$$

The present value of future payoffs when the players form a coalition is

$$PV_C^j = \pi_C^j\left(\frac{1+i}{i+p}\right). \tag{7.4}$$

Finally, the present value of future payoffs from unilateral defection is equal to the net one-time payoff from unilateral defection ($\pi_D{}^j - \pi_N{}^j$) plus the present value of the payoffs when the players do not cooperate, that is

$$PV_D^j = \left(\pi_D^j - \pi_N^j\right) + PV_N^j. \tag{7.5}$$

In general, it will pay for a player to unilaterally defect when the present value of defection is greater than the present value of a coalition, that is

$$\left(\pi_V^j - \pi_N^j\right) + \pi_N^j\left(\frac{1+i}{i+p}\right) > \pi_C^j\left(\frac{1+i}{i+p}\right). \tag{7.6}$$

Inequality (7.6) may be rearranged to yield the following defection rule for finitely-repeated static games with an uncertain end. A player will have a unilateral incentive to defect if the left side of Inequality (7.7) is less than the interest rate used to discount future payoffs. This inequality differs from the defection rule in Inequality (6.8) by the addition of the term $-p(\pi_D{}^j - \pi_N{}^j)$, which captures the uncertainty about the gains from defection. If the left side of this Inequality (7.7) is greater than the interest rate, a player will have an incentive to cooperate. Finally, if the left side is equal to the interest rate, a player will be indifferent between cooperation and defection.

$$\frac{\pi_C^j - \pi_N^j - p\left(\pi_D^j - \pi_N^j\right)}{\pi_D^j - \pi_C^j} < i. \tag{7.7}$$

Solved Exercise

Consider the finitely-repeated, static game depicted in Fig. 6.1. Suppose that the interest rate is 5 percent and the probability that the game will end is 7 percent.

a. *What is the present value of the payoffs when the players do not cooperate?*
b. *What is the present value of the payoffs when the players form a coalition?*
c. *What is the present value of the payoffs from unilateral defection?*
d. *Based on your answers, is a coalition stable?*

e. *Suppose there is an even chance that the game will come to an end in the next stage. For what values of the interest rate will the players have a unilateral incentive to defect?*
f. *Suppose that Inequality (7.7) is satisfied and the interest rate is 5 percent. For what probabilities will the game end in the next stage?*

Solution

a. *The present value of the payoffs when the players do not cooperate is*

$$PV_N^j = \pi_N^j \left(\frac{1+i}{i+p}\right) = 8\left(\frac{1.05}{0.80}\right) = 10.50. \tag{7.8}$$

b. *The present value of the payoffs when the players form a coalition is*

$$PV_C^j = \pi_C^j \left(\frac{1+i}{i+p}\right) = 20\left(\frac{1.05}{0.80}\right) = 26.25. \tag{7.9}$$

c. *The present value of the payoffs from unilateral defection is*

$$PV_D^j = \left(\pi_D^j - \pi_N^j\right) + PV_N^j = (25 - 8) + 10.50 = 27.5. \tag{7.10}$$

d. *Since* $PV_D{}^j > PV_C{}^j$, *the players will have a unilateral incentive to defect.*
e. *From Inequality (7.7) we get*

$$\frac{\pi_C^j - \pi_N^j - p\left(\pi_D^j - \pi_N^j\right)}{\pi_D^j - \pi_C^j} = \frac{12 - 0.5(17)}{5} = 0.7. \tag{7.11}$$

We can expect a coalition to break down for any interest rate greater than 70 percent.
f. *From Inequality (7.7)*

$$\frac{12 - p(17)}{5} < 0.05. \tag{7.12}$$

Solving for p, this game will come to an end for any probability less than 69 percent.

A Word of Caution

Although competition in business is a finitely-repeated game with an uncertain end, caution should be exercised when applying the tools of analysis present in this and the previous chapter. Analyzing strategic behavior is a dynamic process that should reflect the most accurate and available information. Calculating present values requires our best estimate of future payoffs, which should be revised as updated information about earnings and the interest rate becomes available. Another reason for exercising caution has to do with the dynamic nature of business competition. As product lines, customer types, and geographic sales regions change, a manager's strategy choices must change as well.

Practice Exercises

7.1 Analytically, what is the difference between a finitely-repeated game with an uncertain end and an infinitely-repeated game?

7.2 Can a strategy profile that is not a Nash equilibrium in a one-time, static game be a Nash equilibrium strategy if the game is played twice?

7.3 Explain the dynamics of the end-of-game problem.

7.4 A computer software analyst with IBM has given two-week's notice to take a job with Microsoft. The analyst is currently a member of a team that is developing an innovative business application that can be adapted to Microsoft's Excel. Explain how the analyst's job performance might be affected by her decision to resign. Can you think of any other problems for IBM from the analyst's decision to join Microsoft? What advice would you give IBM for handling the analyst's resignation?

7.5 Players A and B are playing a one-time, static game with four strategy profiles: {*A1*, *B1*}, {*A1*, *B2*}, {*A2*, *B1*}, and {*A2*, *B2*}. The corresponding (A, B) payoffs are (5, 5), (2, 6), (6, 2), and (3, 3). Larger payoffs are preferred.

a. What is the Nash equilibrium strategy profile if this game is played one time?
b. What is the Nash equilibrium if the game is played twice?
c. Are there any credible threats that could be made by either player that would result in the best outcome for both players if this game is played three times?

7.6 Consider the static game depicted in Fig. 2.3. Suppose that the interest rate used to discount future payoffs is 5 percent and the probability that the game will end in the next stage is 25 percent.

a. What is the present value of the payoffs if the players do not cooperate?
b. What is the present value of the payoffs if the players form a coalition?
c. What is the present value of the payoffs from unilateral defection? Do the players have a unilateral incentive to defect?
d. Suppose that the probability that the game will end in the next stage is 60 percent. Is a coalition stable?

7.7 Argon Airlines and Boron Airways compete for passengers travelling the Boston–Albany–Buffalo corridor. Both airlines are considering offering discount air fares during the traditionally slow month of February. The four strategy profiles for this static game are {*Discount*, *Discount*}, {*Discount*, *No discount*}, {*No discount*, *Discount*}, and {*No discount*, *No discount*}. The respective (Argon, Boron) profits in millions of dollars are (−1, −1), (30, −5), (−5, 30), and (5, 5).

a. What is the Nash equilibrium strategy profile if this game is played one time?
b. Suppose that the probability that the game will end in the next stage is 15 percent. Above what interest rate can we expect a coalition to fail?
c. Suppose the interest rate is 10 percent. For what probability that the game will end will the players be indifferent between collaboration and defection?

Notes

1. An interesting example of a finitely-repeated coalition game was trench warfare during World War I. The dominant strategy of each combatant was to use artillery to inflict maximum damage. The sides soon realized that this strategy inevitably led to an equally crippling counter-bombardment. Eventually, the combatants realized that the best outcome occurred from simply demonstrating their destructive prowess, rather than actually attacking.
2. See Appendix B for the derivation of this result.

CHAPTER EIGHT

Evolution Games

Introduction

Throughout this text we largely assumed that the existence of a Nash equilibrium depends critically on the assumption of pure rationality, which says that players' strategy choices are best responses to the strategies adopted by rivals. Even its most ardent proponents have come to recognize that the assumption of purely rational behavior is not always satisfied. In some cases, players may be guided less by reason than by instinct, which is the product of repeated responses to specific external stimuli over several generations.

Evolutionary Game Theory

Beginning with Charles Darwin (1859), theories of evolution have provided insights into the physical and mental development of all living creatures. John Maynard Smith's (1982) pioneering work in **evolutionary game theory** explained animal behavior in terms of instinctual strategies that are genetically passed along from generation to generation. New survival strategies arise from mutations. If unsuccessful, survival rates decline and the mutation is not passed along to future generations. If successful, they will come to define a species' behavior.

Reproductive Success

The population-dynamic aspect of evolutionary game theory is significant because it does not depend on a player's purely rational choices, but on the much weaker assumption of **reproductive success** in which successful strategies replace unsuccessful strategies over time. The idea of reproductive success can be traced to the application of game theory to problems in evolutionary biology (see John Maynard Smith, 1982; Josef Hofbauer & Karl Sigmund, 1998).

Animals in the wild are frequently involved in situations involving strategy choices. Stephen and Melinda Pruett-Jones (1994), for example, applied the principles of game theory to analyze the nesting behavior of bowerbirds. Male bowerbirds construct elaborate nests, or bowers, to attract female bowerbirds. Once built, the male bowerbird leaves his nest to destroy neighboring bowers. While marauding, however, his nest is destroyed. The time devoted to destroying and repairing bowers would be better spent foraging and mating. So, what explains this strange behavior?

In evolutionary game theory, animal behavior is "programmed" into its DNA. Genes that comprise strands of DNA "compete" in the sense that they are passed along to future generations in varying proportions. Animals that are successful in the battle for survival tend to have more offspring. As a result, the proportion of animals with genes encoded with a genetically superior survival strategy will come to dominate the population. An evolutionarily stable strategy is one in which the population is resistant to genes encoded with a mutant strategy.[1]

Evolutionary Equilibrium

The Pruett-Joneses (1994) explain the behavior of male bowerbirds by assuming two strategies: *Maraud* and *guard*. A male that marauds spends all his time destroying the nests of rivals. A male that guards spends all his time protecting his nest. The reproductive success of a male bowerbird depends not only on his strategy, but on the strategy adopted by rival bowerbirds. The Pruett-Joneses found that *maraud* strictly dominates *guard*, even though it is in the best interest of male bowerbirds to guard their nests. Since it is not possible for birds to form a coalition (with the possible exception of the Alfred Hitchcock film classic, *The Birds*), this behavior is passed along to future generations until marauding bowerbirds dominate the population. Marauding by bowerbirds constitutes an **evolutionary equilibrium**.

Networks

The ideas underlying the notion of an evolutionary equilibrium have economic and business applications. In business, the word **network** refers to any system of integrated interconnections having a common technical platform. The telecommunications industry, for example, provides voice and data services using the Internet and the World Wide Web. Airlines, railroads, and shipping companies transport people and products using a hub-and-spoke network. The transportation sector provides a platform for networks of delivery services, such as the U.S. Postal Service and FedEx. Financial sector networks include on-site securities exchanges, such as the New York Stock Exchange, electronic over-the-counter markets, such as NASDAQ, B2B (business-to-business) and B2C (business-to-consumer) exchanges, credit and debit card networks, and automated banking networks, such as automated teller machines (ATMs). Many of the features of traditional networks can also apply to **virtual net-**

works, in which compatible goods and services share a common technical platform, such as computers, servers, and software applications geared to Windows operating system.

Firms and households that are part of, utilize, or are organized around a network may enjoy lower cost or receive greater benefits than using different or segregated technologies. Businesses that pool resources, manpower, supplies, and facilities to avoid redundancies, for example, may be able to reduce or eliminate fixed costs, exploit economies of scale, and avoid diseconomies of scale. In a *unidirectional (one-way) network*, for example, data (goods, services and information) flow in one direction. Examples include public utilities that provide electricity, gas, or water to businesses and households. These networks generate significant scale economies and help explain why companies such as Con Edison have an exclusive government franchise in specific geographic regions.

In a **reciprocal (two-way) network**, data flow in two directions. Consider, for example, the hub-and-spoke network depicted in Fig. 8.1. In the telecommunications industry, for example, users A and B are connected by a switching station at the center (hub) of the network.

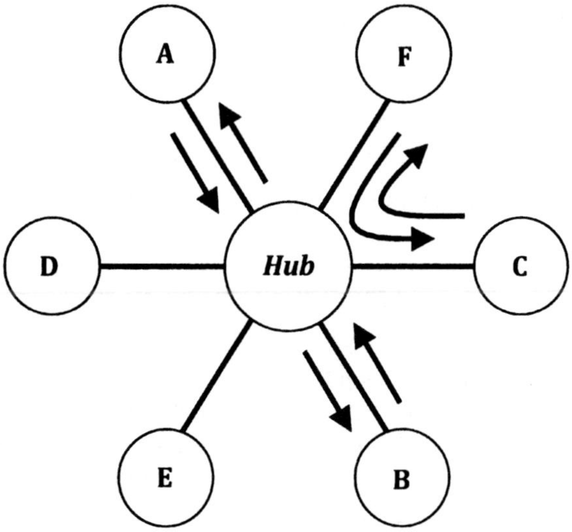

Fig. 8.1.

Another example of a hub-and-spoke network is the airline industry, which links hundreds of cities to provide cost-effective long-distance transportation. Consider a trip from New York City's LaGuardia Airport to Tucson International Airport aboard Delta Airlines. A passenger must first fly along the spoke from New York to Delta's hub in Atlanta, and then along the spoke from Atlanta to Tucson. Other familiar examples of hub-and-spoke networks include parcel and mail delivery by the U.S. Postal Service, parcel delivery by UPS, DHL, and FedEx, and physical check clearing and electronic funds transfer operations performed by the Federal Reserve System.

Positive Feedback Effects

An interesting economic application of reproductive success involves two-way networks, which exhibit *positive feedback effects*, also known as **positive network externalities**. An email or telephone network with only one user is worthless, but becomes exponentially more valuable as the number of users increases. Consider, for example, *Facebook*.[2]

Facebook is a social networking platform that allows its members to create an online nexus of interpersonal relationships. Members of social networks are linked by shared interests, experiences, and backgrounds. *Facebook* makes it possible for different social networks to expand and subsume each other, mush like the growth of separate colonies of bacteria in a Petri dish. In a similar fashion, the benefits to individual members of a social network increase exponentially with its viral-like growth. In the limit, the resulting evolutionary equilibrium is a single, all-inclusive social network that dominates the population.

A numerical example illustrates how positive network externalities can lead to an evolutionary equilibrium. Suppose that the value of membership in a network is proportional to the number of members according to the expression $n(n - 1) = n^2 - n$. The value of a network with a single member is $1^2 - 1 = 0$. The value of the network with two members is $2^2 - 2 = 2$. The value of the network when $n = 10$ is $10^2 - 10 = 90$. If membership increases tenfold to 100 members, the value of the network explodes to $100^2 - 100 = 9{,}900$, or a 110-fold increase! The greater-than-proportional increase in a network's value lures new members until a single dominant network emerges.

Does an evolutionary equilibrium only exist when all players belong to the same network? Not necessarily. For one thing, different networks must be near-perfect substitutes for each other. Moreover, it must not be possible for members of one network to interface with members of another network. At one time, positive feedback effects existed for e-mail networks because members of different networks could not communicate with each other. This ended when software was developed that made communication across different e-mail networks possible. As a result, an evolutionary equilibrium was replaced with multiple network equilibria.

Another example of multiple network equilibria is the competition between broadband technologies. At present, there are two main broadband technologies in use by American households and businesses. Digital subscriber lines (DSL) utilize a copper wire infrastructure that was originally developed to provide home and office telephone service. A coaxial cable infrastructure was originally created to provide pay television service.

DSL and cable provide similar Internet services, including teleconferencing, interactive entertainment, and distance learning. Will competition eventually lead to a dominant Internet technology, or will multiple equilibria emerge consisting of several, or totally different, broadband technologies?

For a single technology to dominate the market, certain conditions must be satisfied, such as economies of scale, positive feedback effects, or a strong consumer preference for a particular technology. While economies of scale appear to exist in satellite and cable broadband distribution, it does not appear that either broadband technology has a distinct competitive advantage. Moreover, consumers do not appear to have a strong consumer preference for a particular broadband technology. Finally, significant positive feedback effects do not exist with any of these broadband technologies.

Although neither DSL nor cable has a distinct competitive advantage at present, the demand for broadband services is sensitive to advances in information technology. While an evolutionary equilibrium in broadband technology does not currently exist, there is no guarantee that this situation will prevail into the future.

Network Game

Positive feedback effects make it difficult for challengers to undermine the competitive advantage of first movers, even when the new network has a superior technology, especially when the incumbent network has a large number of users. This can be illustrated using the two-way network depicted in Fig. 8.2 involving users A and B and a single network provider, H_1. Suppose that the value of this two-way network is $n(n-1) = 2^2 - 2 = 2$.

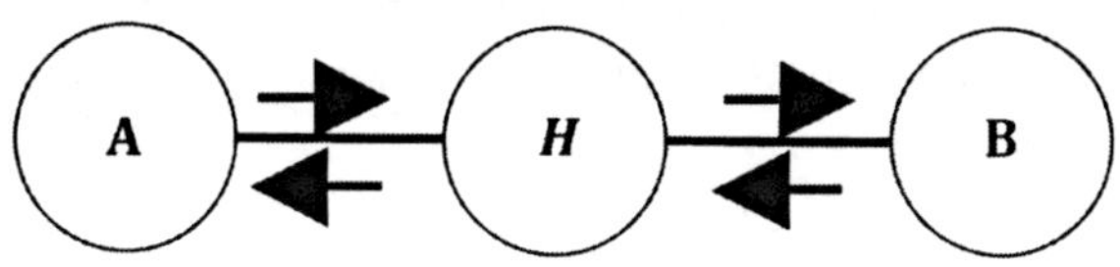

Fig. 8.2.

Suppose that a second network provider with a technology that is twice as efficient enters the market. The value of network H_2 to A and B is $2n(n-1) = 4$. The two networks are exclusive in the sense that they do not communicate with each other. A and B must use either H_1 or H_2. This new situation is depicted in Fig. 8.3.

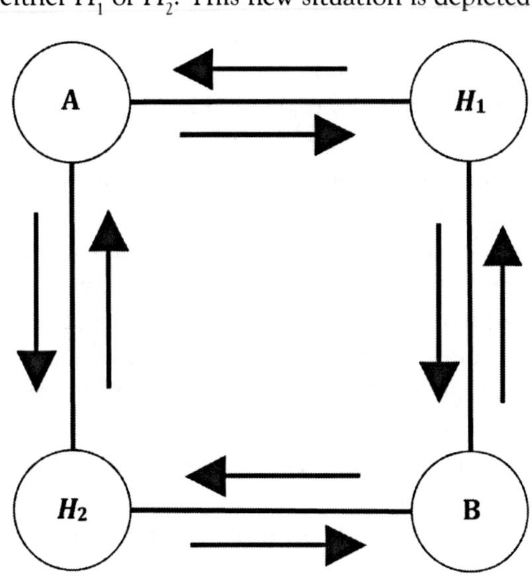

Fig. 8.3.

The reason why it is difficult for the new network to enter the market is depicted in Fig. 8.4. This one-time, static game has two Nash equilibrium strategy profiles: {H_1, H_1} and {H_2, H_2}. The network users are initially at {H_1, H_1}, each receiving value of 2. Although H_2 is superior, neither user has an incentive to switch.

		B: H_1	B: H_2
A	H_1	**(2, 2)**	(0, 0)
	H_2	(0, 0)	**(4, 4)**

Payoffs: (A, B)

Fig. 8.4.

Does the situation depicted in Fig. 8.4 suggest that a first-mover network, such as *Facebook*, will enjoy its monopoly status in perpetuity? Although it is difficult to dislodge an incumbent network, evolutionary game theory suggests that the status-quo will continue until a more successful mutation is introduced into the population. The emergence of *Facebook* usurped instant messaging. What new network technology will replace *Facebook*?

Implications

Positive feedback effects have several interesting implications for market structure and pricing, including a reversal of the law of demand, economies of scale, winner-take-all effects, and increased industrial concentration (Economides, 2006; Liebowitz, 2006).

Positive feedback effects may result in a *reversal of the law of demand,* which states that the quantity demanded of a good or service is inversely related to its price. In the presence of positive network externalities the reverse may be true. Network users may actually be willing to pay more for an additional unit of service given the exponentially increasing value of a larger network. At some price a network demand curve may begin to slope upward.

Economies of scale exist when per unit costs decline as the scale of a firm's operations increases. This phenomenon is typically associated with manufacturing industries that produce goods requiring substantial physical capital outlays. An example of this is an automobile company that incurs substantial start-up costs to design a new model, and to create dies and build assembly facilities for its production.

Research-intensive high-technology industries that form the backbone of the Internet also exhibit significant economies of scale. While the research and development cost of developing computer software can be substantial, the marginal cost of duplicating, servicing, handling, and delivering the product to consumers is relatively insignificant. The impact of economies of scale on the growth of networks is similar, and may be more important, than positive feedback effects.

Positive network externalities and economies of scale combine to give large, incumbent firms a significant competitive advantage over smaller firms and startup companies attempting to penetrate the market. Larger networks have greater positive feedback effects than do smaller networks. Other things being equal, consumers will be willing to pay more to become members of the larger network, which will enhance the large company's profits and competitive edge. This advantage is reinforced by economies of scale that enable large firms to produce at lower per unit cost.

The combined effects of positive network externalities and economies of scale can result in a *winner-take-all* outcome in which a single firm dominates the market, as would seem to be the case with, for example, *Facebook*. The winner-take-all effect can give a pioneering company a significant and insurmountable first-mover advantage. This is especially pronounced in industries where competing firms provide nearly identical goods and services. This outcome is less certain when competing firms that produce differentiate products segment the market, such as the competition between DSL and cable in the market for broadband technology.

Easy entry by new firms leads to increased competition, lower prices, and a tendency to zero economic profits in the long run. This does not appear to be the case with network industries. Even when there are low barriers to entry, significant network feedback effects and economies of scale tend to lead to *increased industrial concentration*. In general, firms in highly concentrated industries have market power, with the ability to charge prices that are greater than marginal cost to earn above-normal returns in the long run. This presents a challenge to government anti-trust regulators whose mandate is to protect consumers by promoting competition. Because the winner-take-all effect is characteristic of network industries, superimposing a competitive market structure may be futile at best, counterproductive at worst.

Practice Exercises

8.1 Explain the concept of an evolutionary equilibrium.

8.2 Distinguish between reproductive success and pure rationality.

8.3 How have game theorists applied an evolutionary stable strategy to explain changes in social attitudes regarding sexual preference, marriage, adultery, and age, race, and gender discrimination?

8.4 QWERTY keyboards were invented in 1878 to minimize typewriter keys from jamming. This arrangement of letters has long been recognized as inefficient and confusing. Although jamming is no longer an issue with modern computer keyboards, QWERTY continues to dominate the market.

a. Would you consider QWERTY a network?
b. Why has QWERTY been resistant to a more efficient arrangement of letters on a keyboard?

8.5 Linux is generally considered to be superior to Microsoft's Windows and Apple's Macintosh operating systems.
a. Do Windows and Macintosh constitute networks?
b. Does this explain why Linux has been unable to establish a significant foothold in the computer software market?

8.6 The Internet auction site eBay is an example a hub-and-spoke network that connects multiple buyers and sellers. Explain how first-mover and positive feedback effects enabled eBay to dominate the on-line auction market.

8.7 How would you describe local cable television networks? What effect has this kind of network had on prices paid by cable television subscribers?

8.8 The Pruett-Joneses have argued that marauding is an evolutionary stable strategy among bowerbirds. Suppose initially that the population is evenly divided between *guarding* and *marauding* bowerbirds. There are four strategy profiles: {*Guard*, *Guard*}, {*Guard*, *Maraud*}, {*Maraud*, *Guard*}, and {*Maraud*, *Maraud*}. The respective payoffs are (2, 2), (3, 0), (0, 3), and (1, 1). What is the evolutionary equilibrium for this game?

8.9 *Twitter* is a real-time, on-line communication service that allows users to simultaneously share information with every other user. Is *Twitter* a network, a bulletin board, or both? Diagram and explain.

Notes

1. Formally, strategy s^* is an evolutionary stable strategy for every other strategy $s' \neq s^*$ if $u(s^*, s^*) \geq u(s', s^*)$ and $u(s', s^*) = u(s', s^*)$, such that $u(s^*, s') > u(s', s')$.

2. In 2004 Mark Zuckerberg and fellow Harvard students Chris Hughes, Andrew McCollum, Dustin Moskovitz, and Eduardo Saverin launched the social networking website Facebook. Membership was originally limited to students at Harvard, but quickly expanded to include students at other Ivy League colleges and Stanford. Eight years later, membership in Facebook exceeded 800 million users. Today, the company has an estimated market value of \$80–\$100 billion and a payroll of 3,000 employees. What accounted for the company's viral growth?

CHAPTER NINE

Tit-for-Tat

Introduction

Many cooperative arrangements in static coalition games have survived despite the fact that the players had an incentive to defect. In some cases, coalitions that broke down were subsequently reformed, even though the conditions that led to their disintegration still existed. One possible explanation is the use of a trigger strategy that punishes defection, but that does not rule out future cooperation. One such trigger strategy is tit-for-tat.

Tit-for-Tat

Tit-for-tat is a variation of the eye-for-an-eye rule found in the Bible (Exodus 21:23–24, *King James Version*) to enforce cooperation in finitely-repeated, static coalition games. A player who adopts a **tit-for-tat** strategy will cooperate if the other player cooperates, but will defect if the other player defects.

According to Robert Axelrod (1984, p. 20), tit-for-tat can stabilize coalitions in finitely-repeated, games when the players "have a sufficiently large chance to meet again so that they have a stake in their future cooperation." Tit-for-tat is an effective trigger strategy because it is a "combination of being nice, retaliatory, forgiving, and clear" (p. 54). Tit-for-tat is nice because it never initiates defection. It is retaliatory because it will never let a defection go unpunished. It is forgiving because it does not rule out the possibility of future cooperation. And, it is clear because it is simple and easy to understand.

The most severe shortcoming of a tit-for-tat strategy is the possibility that a player will misinterpret a rival's actions or intentions. This miscalculation may take one of two forms. Either a player misinterprets a rival's move, or a player correctly interprets a rival's moves, but misinterprets a rival's motives. Such miscalculations can lead to a chain reaction of move and countermove that results in a complete breakdown of the coalition. At no point does tit-for-tat accept defection without retaliation. Moreover, there is no mechanism for short-circuiting this downward spiral once the process has begun.

To illustrate tit-for-tat, suppose that firms A and B sell nearly-identical products for which there are no close substitutes. Each firm has a core of loyal customers. There is also a third group of customers that will always buy from the firm that charges the lowest price. Either company that unilaterally charges a slightly lower price will capture this price-sensitive group and realize a substantial increase in its operating profits.

Suppose initially that both firms agree to charge the same monopoly price and split sales to the price-sensitive group. To maintain price discipline, each firm adopts the tit-for-tat strategy that if the other firm cuts its price, it will respond by cutting price by a little more. Since neither firm wants to lose money, the lowest price charged equals marginal cost.

Now, suppose that A mistakenly believes that B is planning to shave its price. Firm A preempts this move by cutting its price by a little more. Firm B, which had no intention of breaking the peace, retaliates by lowering its price, and so on until the price-fixing agreement completely unravels until each firm earns a zero economic profit.

The reason why tit-for-tat is so compelling is that it does not rule out the possibility of the process working in reverse. Suppose that A realizes that it misinterpreted B's intentions when it decided to cut its price, thereby making both firms worse off. Recognizing its mistake, A tests the water by raising its price by a small amount. B responds by raising its price. These tentative first steps are taken as a signal that both firms would be better off by returning to the *status quo ante*. In fact, mutual recognition that price competition is counterproductive may be enough to keep either firm from starting a war it cannot win.

Another reason why tit-for-tat is frequently observed in the real world is that it performs well against competing trigger strategies. In an interesting study of finitely-repeated, coalition games, Robert Axelrod (1984) invited several game theorists to participate in a computer tournament in which they were asked to submit competing one-on-one strategies in which players could cooperate or defect. Although tit-for-tat never beat, and sometimes tied, competing enforcement strategies, it won the tournament by accumulating the highest overall score.

An interesting real-world example of tit-for-tat occurred in the 1990's involving Philip Morris and British American Tobacco (BAT), which dominated the Costa Rican cigarette market. At the time, Philip Morris catered to the premium and mid-priced segments of the market with its Marlboro and Derby brands. BAT focused on the low-priced segment with its Delta brand.

Storm clouds began to form in the early 1990s as the health dangers of smoking became more widely known. The effect in Costa Rica was a greater reduction in the demand for premium and mid-priced cigarettes than for low-priced brands. For the first time since the early 1980s, BAT's market share began to increase.

In early 1993, Philip Morris responded to the decline in its market share by cutting prices 40 percent. BAT retaliated by slashing its prices by 50 percent, which was slightly above its marginal cost. This price war lasted two years. Although industry sales increased by 17 percent because of generally lower prices, market shares of Philip Morris and BAT remained largely unchanged. In the meantime, Philip Morris and BAT profits plummeted \$8 million and \$20 million, respectively.

What prompted Philip Morris to initiate the price war? One explanation is that Philip Morris had successfully used this strategy in other Central American countries, which underscores the importance of correctly anticipating a rival's countermove. A rapid, retaliatory strike can serve as an effective enforcement mechanism to keep rivals from using price cuts to increase its market share.

End-of-Game Problem

In its simplest form, players will adopt, and continue to play, tit-for-tat even when the end of the game is certain. But is this realistic in light of our earlier discussion of the *end-of-game problem* in which the use of trigger strategies to enforce cooperation will fail because retaliation is not possible beyond the last stage? This effectively transforms each stage into a one-time game. How the players move in the last stage determines how they will move in the first, and all subsequent, stages.

Even more provocative is why a player would adopt a tit-for-tat strategy in the first place. A player in a coalition game is unlikely to play tit-for-tat unless there is a reasonably high expectation that the other player will do so as well since no one wants to get stuck with the sucker's payoff. So, what is it about tit-for-tat that promotes cooperation, and how does it keep finitely-repeated static games with a certain end from unraveling into a series of one-time, static games?

David L. Kreps, Paul Milgrom, John Roberts, and Robert Wilson (1982) answered this question by demonstrating that under the right conditions tit-for-tat is not only a best response during the early stages of a coalition game, but that it largely overcomes the end-of-game problem. They accomplished this by making certain assumptions about what the players know about each other, and how these beliefs might change during the game.

Kreps, et al., began by recognizing that tit-for-tat is characterized by incomplete (asymmetric) information. That is, each player is certain about his or her moves, but can only infer the intentions of rivals. On the other hand, if either player defects at any point in the game, neither player will cooperate thereafter. Finally, regardless of what the players decide in the earlier stages, each has an incentive to defect in the last stage. In their description of tit-for-tat, all undiscounted payoffs are realized at the end of the game.

A detailed description of the Kreps, et al. analysis is beyond the scope of this book.[1] This section merely summarizes the main conclusions of their analysis. Consider the coalition game depicted in Fig. 9.1 in which A and B must decide whether to cooperate (C) or defect (D), where $a_i > b_i$, $d_i > a_i$, and $b_i > c_i$, where i = A or B.[2] The reader should confirm that the Nash equilibrium strategy profile for this one-time static game is {D, D}, although it is in both players' best interest to form a coalition and adopt the strategy profile {C, C}. In this one-time game, cooperation is unstable since both players have an incentive to defect.

		B: C	B: D
A	C	(a_A, a_B)	(c_A, d_B)
	D	(d_A, c_B)	(b_A, b_B)

Payoffs: (A, B)

Fig. 9.1.

Now, suppose the game depicted in Fig. 9.1 is finitely repeated with T > 3 stages. Assume that each player begins by cooperating, and then mimics the other player's move with probability p thereafter. It may be demonstrated that cooperation is a best response through period T – 2, and that defection is a best response in stages T – 1 and T, provided that

$$pa_i + (1 - p)c_i \geq b_i, \tag{9.1}$$

$$a_i + p(d_i - b_i) \geq d_i, \tag{9.2}$$

$$a_i + b_i \leq c_i + d_i. \tag{9.3}$$

For cooperation through period T – 2 to be possible, conditions (9.1) to (9.3) must be satisfied for both players. If these conditions are satisfied for only one of the players, either because of differences in player beliefs, payoffs, or both, cooperation will break down after the first stage. When this occurs, the cooperating player will suffer a one-time loss (compared with not cooperating) of $(c_i - b_i)$, while the other player enjoys a one-time gain of $(d_i - b_i)$.

Solved Exercise

Consider the pricing game depicted in Fig. 2.3 in which larger payoffs are preferred. Assume that the firms' undiscounted profits are realized at the end of the game.

a. *What is the Nash equilibrium strategy profile in this one-time, complete information, pure-strategy static game?*

b. *Suppose that this game is repeated 10 times. If larger payoffs are preferred, what is the Nash equilibrium if this game is characterized by complete information? What is each firm's total profit?*
c. *Suppose that each firm is considering charging a high price. Each believes there is a 50 percent chance that the other will adopt a tit-for-tat strategy. What are the firms' expected undiscounted, end-of-game profits?*
d. *What are the firms' expected, undiscounted, end-of-game profits if each believes that there is a 10 percent chance that the other will play tit-for-tat?*

Solution

a. *This is an example of a one-time coalition game in which it is in both firms' best interest to cooperate by charging a high price, but where each has an incentive to adopt its strictly-dominant low price strategy. The resulting strategy profile is {Low price, Low price}, with each firm realizing a profit of 0.25, or $250,000.*
b. *In a game of complete information, the end-of-game problem suggests that each player will adopt a low-price strategy in each stage. Tit-for-tat will not bind the firms to a high-price coalition. Each firm will realize an end-of-game total profit of 2.5, or $2.5 million.*
c. *Substituting this information into conditions (9.1) to (9.3) suggests that a tit-for-tat strategy will successfully bind the coalition for the first eight stages, but that each will defect in the ninth and tenth stages. The expected end-of-game profit for each firm is 8(1) + 2(.25) = 8.5, or $8.5 million.*
d. *Although condition (9.3) is satisfied, conditions (9.1) and (9.2) are not. Tit-for-tat will not bind the players to this coalition and the expected end-of-game total profit for each firm will be the same as if this game were characterized by complete information.*

Practice Exercises

9.1 What is it about tit-for-tat that makes it an effective trigger strategy? Why did tit-for-tat fail to maintain price stability in the Costa Rican cigarette market in the early-1990s?

9.2 What is the main weakness of a tit-for-tat trigger strategy?

9.3 Suppose that British Airways and Delta Air Lines are competing for business customers on the lucrative New York to Paris route during the off-peak season. Both would like to lure customers by offering lower fares. There are four pricing-strategy profiles: {$750, $750}, {$750, $500}, {$500, $750}, and {$500, $500}. The undiscounted (British Airways, Delta Air Lines) payoffs are (2, 2), (0, 3), (3, 0), and (1, 1), which are realized at the end of the game.

a. If larger payoffs are preferred, what is the Nash equilibrium strategy profile in this complete-information, one-time static game? What is the payoff to each airline?
b. Suppose that this game is repeated 12 times. Assuming complete information, what are the undiscounted, end-of-game payoffs?

c. Suppose that this game is repeated 12 times and each airline believes that there is a 75 percent chance that the other airline will cooperate. What are the expected undiscounted, end-of-game payoffs for each airline?
d. Suppose that each airline believes that there is a 40 percent chance that the other airline will cooperate. What are the expected, undiscounted, end-of-game payoffs?
e. Suppose that British Airways believes that there is a 75 percent chance that Delta will cooperate, while Delta believes that there is a 40 percent chance that British Airways will cooperate. What are the airlines' expected, undiscounted end-of-game payoffs?

Notes

1. A simplified version of this analysis may be found in Robert Gibbons (1992).
2. In the classic prisoner's dilemma coalition game, it is typically assumed that $a_A = a_B$, $b_A = b_B$, $c_A = c_B$, and $d_A = d_B$. We also assume that $c_i + d_i < a_i + a_i$, which guarantees that it is in both players' best interest to cooperate until the last stage rather than alternating between cooperation and defection.

CHAPTER TEN

~

Mixing Pure Strategies

Introduction

We have learned that identifying unique Nash equilibria in pure-strategy, static games in which a player does not have a dominant strategy can be problematic. Such games may have multiple Nash equilibria, or may have none at all. Game theory would long ago have lost its intellectual appeal if its applications were limited to situations involving dominant pure strategies. Game theory is a powerful tool for analyzing strategic situations precisely because a unique Nash equilibrium exists for all static games, although not always in pure strategies.

Recall that a pure strategy is a complete and nonrandom game plan. By contrast, a mixed strategy involves randomly mixing pure strategies. We will begin our discussion of mixed strategies by examining pure-strategy, static competition games with zero-sum payoffs.

Zero-Sum Games

Nearly a quarter of a century before John Nash demonstrated the existence of a "fixed point" equilibrium, John von Neumann was breaking new ground in the analysis of strategic behavior. Although not a very good player himself, von Neumann's inspiration was poker, which is an example of a zero-sum game in which one player's gain is another player's loss.

An important aspect of poker games is the bluff, which occurs when a player bets big on a bad hand. The objective of bluffing is to get opponents to "fold" (quit) by persuading them that they are sitting on losing hands. Von Neumann found that for this strategy to be successful, players must randomly bet big on both good and bad hands. Rivals will see through a player who repeatedly bets big on inferior hands.

When this happens, not only will bluffing become ineffective, but rivals will exploit the attempt. This observation led von Neumann (1928) to publish the proof of his famous *minimax theorem*.

Matching Pennies

To illustrate the concept of a zero-sum game, consider the matching pennies game in which Ben and Jerry simultaneously show a penny with heads (obverse) up or tails (reverse) up. If the pennies match (two heads or two tails), Ben wins Jerry's penny. If the sides do not match, Jerry wins Ben's penny. The normal form of the matching pennies game is depicted in Fig. 10.1, which is zero sum because Ben's gain is Jerry's loss, and vice versa. In this game, neither player has a dominant strategy, nor is it possible to identify a pure-strategy Nash equilibrium by eliminating dominated strategies.

		Jerry	
		Heads	*Tails*
Ben	*Heads*	(**1**, −1)	(−1, **1**)
	Tails	(−1, **1**)	(**1**, −1)

Payoffs: (Ben, Jerry)

Fig. 10.1.

Minimax Theorem

The first formal proof of the **minimax theorem** was presented by John von Neumann in a paper delivered to the Gottingen Mathematical Society on December 7, 1926. Von Neumann demonstrated that two rational players in zero-sum games will attempt to both maximize their gains and minimize their losses. John Forbes Nash later extended von Neumann's analysis to include non-zero-sum games. It was subsequently demonstrated that maximin and Nash equilibrium strategies are one and the same.

To illustrate the minimax theorem, suppose that two players submit one-time bids for the distribution of $10. If the sum of the bids is less than or equal to $10, the players receive their proposed bids. If the sum of the bids exceeds $10 both players receive nothing. The Nash equilibrium strategy profile is for both players to submit bids of $5. No other strategy simultaneously maximizes the players' gain and minimizes their losses.

Solved Exercise

Consider the one-time, zero-sum, static game involving players A and B. There are four strategy profiles: {A1, B1}, {A1, B2}, {A2, B1}, and {A2, B2}. The respective payoffs are (1, –1), (2, –2), (–2, 2), and (–1, 1).

a. *What is the strategy profile if both players adopt a minimax strategy?*
b. *Is this strategy profile a Nash equilibrium?*

Solution

a. *If A moves A1, B's maximum payoff is –1. If A moves A2, B's maximum payoff is 2. Player A should move A1 to minimize B's maximum payoff. If A moves A1, A's minimum payoff is 1. If A moves A2, A's minimum payoff is –2. A1 maximizes A's minimum payoff. Since A1 both minimizes B's maximum payoff and maximizes A's minimum payoff, A1 is A's minimax strategy. Since the payoffs are symmetrical, B's minimax is B1. The resulting strategy profile is {A1, B1}.*
b. *The strategy profile {A1, B1} is also a Nash equilibrium because it represents each player's best response to the strategy adopted by the other player.*

Mixed Strategies

One of the problems with the minimax theorem is that it does not apply equally to all zero-sum games. To see this, consider the matching pennies game in Fig. 10.1. This game does not have a unique, pure-strategy Nash equilibrium because the minimum payoff for each continuously distributed strategy profile is the same. Fortunately, this is not the end of the story. By randomizing strategies, we can transform a static game in two pure strategies into a static game with an infinite number of strategy profiles. Randomizing pure strategies is referred to as a *mixed strategy*. This collection of randomized pure strategies is inclusive in that the sum of the probabilities is unity.

Although the matching pennies game does not have a pure-strategy Nash equilibrium, it does have a unique, mixed-strategy Nash equilibrium provided that Ben and Jerry randomize their choice of *heads* and *tails* with equal probabilities of 0.5, such as by flipping their fair coins. The reason that this is a unique Nash equilibrium is because neither player can obtain a better payoff by randomly mixing heads and tails in any other way. To see this, suppose that Ben randomly shows *heads* 20 percent and *tails* 80 percent. Recall that Jerry wins if the sides do not match. Thus, it will be in Jerry's best interest not to randomize strategies, but to adopt a pure *heads* strategy. Jerry's expected payoff is $0.2(-1) + 0.8(1) = 0.6$ cents and Ben's expected payoff is $0.2(1) + 0.8(-1) = -0.6$ cents.

Is there any other mixing rule that will improve Ben's payoff? If so, then Ben's mixed strategy of 20 percent *heads* and 80 percent *tails* is not a Nash equilibrium. Suppose that Ben adopts a 40 percent *heads* and 60 percent *tails* mixed strategy. Jerry's expected payoff from a pure *heads* strategy is $0.4(-1) + 0.6(1) = 0.2$ cents and Ben's expected payoff $0.4(1) + 0.6(-1) = -0.2$ cents. Ben's payoff is algebraically greater, but he still suffers an expected loss.

Suppose that Ben randomizes *heads* and *tails* with equal probabilities of 0.5. Ben's expected payoff is $0.5(-1) + 0.5(1) = 0$ cents. A 50-50 mixing rule is strictly dominant since no other strategy will result in a better payoff, regardless of the strategy adopted by Jerry. Because of the symmetrical payoffs, Jerry's optimal mixing rule is 50-50 as well. Although the matching pennies game does not have a unique, pure-strategy Nash equilibrium, by eliminating dominated strategies (mixing rules), the result is a unique **mixed-strategy Nash equilibrium**.

Solved Exercise

Consider the following static game in which management wants labor to work hard, but labor has an incentive to goof off. Management can monitor or ignore workers on the job. Labor can either work or shirk. Management loses when it monitors and labor works because it incurs unnecessary monitoring costs; labor wins because it is recognized and rewarded for hard work (–1, 1). Management wins when it ignores when labor works because it does not incur monitoring costs; labor loses because work is not recognized (1, –1). Management wins when it monitors shirking workers who are disciplined (1, –1). Finally, management loses when it ignores shirking workers who adversely affect company performance (–1, 1).

a. *Does this game have a pure-strategy Nash equilibrium?*
b. *What is the mixed-strategy Nash equilibrium? Explain.*

Solution

a. *This game does not have a pure strategy Nash equilibrium. The best response by each player changes when the strategy of the other player changes. If management monitors, labor's best response is to work. If management ignores, labor's best response is to shirk.*
b. *The mixed-strategy Nash equilibrium for both players is a randomized 50-50 mixing rule. Although the expected payoff for each player is zero, no other mixing rule will result in a better payoff.*

Optimal Mixing Rules

We saw in the matching pennies game that it may be possible for a player to improve his payoff by randomly mixing pure strategies. Being unpredictable, however, does not always mean flipping a fair coin. That is, 50-50 is not always an optimal mixing rule. To see this, consider the game depicted in Fig. 10.2 in which the New York Giants are trying to gain first-down yardage against the Baltimore Ravens.

		Ravens	
		Pass defense	*Run defense*
Giants	*Pass offense*	(10, 90)	(80, 20)
	Run offense	(70, 30)	(40, 60)

Success rates: (Giants, Ravens)

Fig. 10.2.

The payoffs represent the probabilities that each side will be successful from any combination of offensive and defensive plays. If the Giants correctly anticipate that the Ravens will defend against the run, for example, there is an 80 percent chance that they will gain first-down yardage by passing the ball and a 20 percent chance that the Ravens will prevent a first down. On the other hand, if the Giants incorrectly predict that the Ravens will defend against the run, there is only a 10 percent chance that they will get a first down by passing the ball and the probability that the Ravens will stop the Giants jumps to 90 percent. The game in Fig. 10.2 does not have a Nash equilibrium in pure strategies.

To determine whether there exists a Nash equilibrium in mixed strategies, we will first consider this game from the Ravens' perspective. As we move from left to right along the horizontal axis in Fig. 10.3, the probability that the Ravens adopt a *pass defense* increases from zero to 100 percent as the probability that they adopt a *run defense* simultaneously decreases from 100 to zero percent. Suppose that the Ravens adopt a pure *pass defense*. There is a 10 percent probability that the Giants will gain first-down yardage with a pure *pass offense*. If the Ravens adopt a pure *run defense* strategy, there is an 80 percent probability of a first down with a pure *pass offense*.

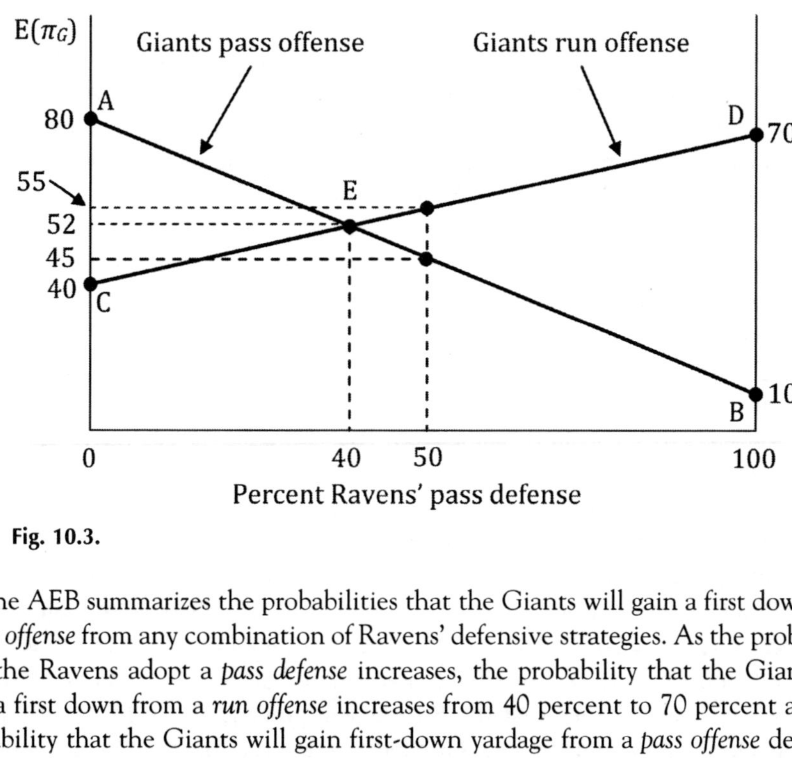

Fig. 10.3.

Line AEB summarizes the probabilities that the Giants will gain a first down with a *pass offense* from any combination of Ravens' defensive strategies. As the probability that the Ravens adopt a *pass defense* increases, the probability that the Giants will gain a first down from a *run offense* increases from 40 percent to 70 percent and the probability that the Giants will gain first-down yardage from a *pass offense* decreases from 80 percent to 10 percent. How should the Ravens randomly mix their plays to simultaneously maximize their chances of stopping the Giants while minimizing the Giants' chances of getting a first down?

If the Ravens adopt a pure *pass defense* strategy, there is a 70 percent chance that the Giants will get a first down with a *run offense*. If the Ravens adopt a pure *run defense* strategy, there is a 40 percent probability that the Giants will gain first-down yardage with a *run offense*. Line CED in Fig. 10.3 summarizes the probabilities that the Giants will successfully gain a first down with a *run offense* strategy from any combination of Ravens' defensive strategies. How should the Ravens randomize its defensive plays to minimize the Giants' chances of gaining first-down yardage?

Suppose that the Ravens' head coach randomizes plays by flipping a fair coin, as in the matching pennies game. By adopting a pure *pass offense* strategy, the expected probability that the Giants will gain first down yardage is 0.5(80) + 0.5(10) = 45 percent, which means that the probability that the Ravens will stop the Giants is 55 percent. On the other hand, if the Giants adopt a pure *run offense*, the probability of a first down is 0.5(70) + 0.5(40) = 55 percent of the time, which means that there is a 45 percent chance that the Ravens will prevent a first down. Thus, if the Ravens use a 50–50 mixing rule, it will be in the Giants' best interest to adopt a pure *run offense* strategy since this will result in a first down 55 out of every 100 offensive plays.

The Ravens 50-50 mixing rule is clearly superior to a pure *run defense* since the Giants probability of getting a first down with a pure *pass offense* is 80 percent. It is also superior to a pure *pass defense* since the probability that the Giants will gain first-down yardage is 70 percent with a pure *run offense*. But, the Ravens do even better.

The Raven's **optimal mixing rule** is a randomized combination of pure strategies that minimizes the Giants' maximum expected chance of making a first down, but also maximizes its own minimum expected chance of stopping the Giants. The Giants' maximum expected payoffs for any combination of Ravens strategies are given by the line AED. To minimize the Giants' maximum expected payoff $E(\pi_G)$, the Ravens should adopt the 40-60 mixing rule, which corresponds to point E in Fig. 10.3.[1] That is, the Ravens should randomly defend against the pass 4 out of 10 times, and defend against the run 6 out of 10 times. The Giants' expected probability of a first down is 52 percent. Point E is called a "saddle point" since it represents the minimum of the Giants' maximum expected payoffs.

Now consider the first-down game from the Giants' perspective. If the Giants adopt a pure *pass offense* there is a 90 percent chance that the Ravens will prevent a first down with a pure *pass defense*. If the Giants adopt a pure *run offense*, there is a 30 percent probability that the Ravens will prevent a first down with a pure *pass defense*. Line HJI in Fig. 10.4 summarizes the probabilities that the Ravens will prevent a first down with a *pass defense* from any combination of Giants' offensive strategies. If the Giants adopt a pure *run offense*, there is a 60 percent probability that the Ravens will prevent a first down with a pure *run defense*. If the Giants adopt a pure *pass offense*, there is a 20 percent chance that the Ravens will prevent a first down with a pure *run defense*. Line FJG summarizes the probabilities that the Ravens will prevent a first down with a *run defense* from any combination of Giants' offensive strategies.

The Giants' optimal mixing rule is a randomized combination of pure strategies that minimizes the Ravens' expected maximum payoffs. The Ravens' maximum payoffs from any combination of Giants' offensive strategies are given by the line FJI. To minimize the Ravens' maximum expected payoff, the Giants should adopt a 30-70 mixing rule. That is, the Giants should randomly pass 3 out of 10 times and randomly run 7 out of 10 times. The Ravens' expected probability of preventing a first down is

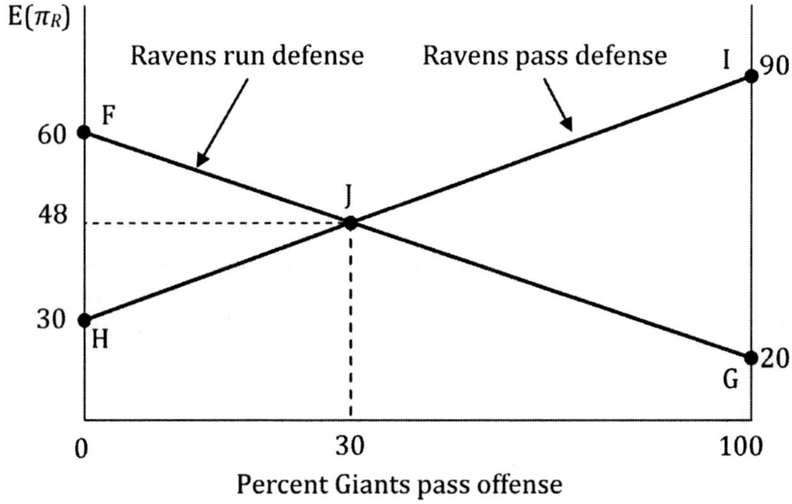

Fig. 10.4.

48 percent. By definition, the sum of the Ravens' and Giants' expected probabilities is 52 + 48 = 100 percent.

Calculating Optimal Mixing Rules

In this section we develop a computationally efficient method of calculating an optimal mixing rules for two-player, static games. Consider the first-down game in Fig. 10.2. Recall that a player's expected payoff depends on a rival's mixing rule. Suppose that α is the probability that the Giants will adopt a *pass offense* and $(1 - \alpha)$ the probability that the Giants will adopt a *run offense*. Let β be the probability that the Ravens will adopt a *pass defense* and $(1 - \beta)$ the probability that the Ravens will adopt a *run defense*.

The Giants' expected probability of a first down from a *pass offense* is

$$E(\pi_G^{PO}) = 10\beta + 80(1-\beta) = 80 - 70\beta. \qquad (10.1)$$

The Giants' expected probability from a *run offense* is

$$E(\pi_G^{RO}) = 70\beta + 40(1-\beta) = 40 + 30\beta. \qquad (10.2)$$

Solving Eqs. (10.1) and (10.2) is the Ravens' optimal mixing rule $\beta = 4/10$ and $(1 - \beta) = 6/10$. Substituting this result into Eq. (10.1) or Eq. (10.2), the Giants' expected probability of gaining first-down yardage is 52 percent. These are precisely the results identified in Fig. 10.3.

Similarly, the Ravens' expected probability of preventing a first down payoff from a *pass defense* is

$$E(\pi_R^{PD}) = 90\alpha + 30(1-\alpha) = 30 + 60\alpha. \qquad (10.3)$$

The Ravens' expected probability from a *run defense* is

$$E(\pi_R^{RD}) = 20\alpha + 60(1-\alpha) = 60 - 40\alpha. \tag{10.4}$$

Solving Eqs. (10.3) and (10.4), the Giants' optimal mixing rule is α = 3/10 and (1 – α) = 7/10. Substituting this result into Eq. (10.3) or Eq. (10.4), the Ravens' expected probability of preventing a first down is 48 percent. These are the same results identified in Fig. 10.4.

The above examples demonstrate that a unique Nash equilibrium exists for any game with a finite number of players and strategies, although mixing pure strategies may be required. This phenomenon is referred to as the **Nash equilibrium existence theorem.**

Solved Exercise

Consider the matching pennies game in Fig. 10.1.

a. *Calculate Ben's and Jerry's optimal mixing rules.*
b. *What are Ben's and Jerry's expected payoffs?*

Solution

a. *Ben's mixing rule is (α, 1 – α), where α is the probability of playing heads and 1 – α is the probability of playing tails. Jerry's mixing rule is (β, 1 – β), where β is the probability of playing heads and 1 – β is the probability of playing tails. Ben's expected payoff from a heads strategy is*

$$E(\pi_B^H) = 1\beta - 1(1-\beta) = -1 + 2\beta. \tag{10.5}$$

Ben's expected payoff from a tails strategy is

$$E(\pi_B^T) = -1\beta + 1(1-\beta) = 1 - 2\beta. \tag{10.6}$$

Solving Eqs. (10.5) and (10.6), β = (1 – β) = 1/2. Jerry's optimal mixing rule is (β, 1 – β) = (1/2, 1/2).
Jerry's expected payoff from a heads strategy is

$$E(\pi_J^H) = -1\beta + 1(1-\alpha) = 1 - 2\alpha. \tag{10.7}$$

Jerry's expected payoff from a tails strategy is

$$E(\pi_J^T) = 1\alpha - 1(1-\alpha) = -1 + 2\beta. \tag{10.8}$$

Solving Eqs. (10.7) and (10.8), α = (1 – α) = 1/2. Ben's optimal mixing rule is (α, 1 – α) = (1/2, 1/2).

b. Ben's expected payoff is

$$E(\pi_B) = -1 + 2\left(\frac{1}{2}\right) = 1 - 2\left(\frac{1}{2}\right) = 0. \tag{10.9}$$

Jerry's expected payoff is

$$E(\pi_J) = 1 - 2\left(\frac{1}{2}\right) = -1 + 2\left(\frac{1}{2}\right) = 0. \tag{10.10}$$

Solved Exercise

Consider the game in Fig. 4.9 in which larger payoffs are preferred.

a. Does this game have a Nash equilibrium in pure strategies?
b. Determine the optimal mixing rule for each player.
c. What are the expected payoffs if each player adopts its optimal mixing rule?

Solution

a. This game has two Nash equilibria in pure strategies: {A1, B1} and {A2, B2}.
b. Let α be the probability that player A moves A1 and (1 – α) that A moves A2. Let β be the probability that player B moves B1 and (1 – β) that B moves B2. Player A's expected payoff from an A1 strategy is

$$E(\pi_A^{A1}) = 100\beta + 120(1-\beta) = 120 + 20\beta. \tag{10.11}$$

Player A's expected payoff from an A2 strategy is

$$E(\pi_A^{A2}) = 50\beta + 150(1-\beta) = 150 - 100\beta. \tag{10.12}$$

Solving Eqs. (10.11) and (10.12), player B's optimal mixing rule is (β, 1 – β) = (3/8, 5/8).
Player B's expected payoff from a B1 strategy is

$$E(\pi_B^{B1}) = 200\alpha + 50(1-\alpha) = 50 + 150\alpha. \tag{10.13}$$

Player B's expected payoff from a B2 strategy is

$$E(\pi_B^{B2}) = 75\alpha + 150(1-\alpha) = 150 - 75\beta. \tag{10.14}$$

Solving Eqs. (10.13) and (10.14), player A's optimal mixing rule is (α, 1 – α) = (4/9, 5/9).

c. *If both players adopt their optimal mixing rules, the expected payoffs are*

$$E(\pi_A) = 120 + 20\left(\frac{3}{8}\right) = 150 - 100\left(\frac{3}{8}\right) = 112.50. \quad (10.15)$$

$$E(\pi_B) = 50 + 150\left(\frac{4}{9}\right) = 150 - 75\left(\frac{4}{9}\right) = 116.67. \quad (10.16)$$

When to Use Optimal Mixing Rules

We saw in the previous section that a player can exploit any deviation from a rival's optimal mixing rule. Suppose, for example, that the Giants in Fig. 10.4 replace its optimal 30-70 mixing rule with a pure *pass offense* strategy. The Ravens will exploit this by adopting a pure *pass defense* strategy. This reduces the Giants' expected probability of a first down from 52 to 10 percent.

What if the Giants deviate from its optimal mixing rule for relatively unimportant plays to lull the Ravens into a *pass defense* to run the ball when it really matters, such as to score a touchdown? Once the Ravens abandon its optimal mixing rule to exploit a perceived deviation by the Giants, they become vulnerable to the "bait and switch." To avoid falling into this trap, the Ravens should not deviate from its optimal mixing rule, unless there is some compelling reason to do so. The same is also true for the Giants. Any deviation will increase a rival's advantage. A mixed strategy Nash equilibrium is only possible when both players adopt their optimal mixing rule.

How to Use Optimal Mixing Rules

It is not enough for a player to use its optimal mixing rule to minimize a rival's maximum payoffs and maximize their own minimum payoff. To be effective, these mixing strategies must be randomized. It is important, for example, for the Giants to run the ball in all third-down situations since the Ravens will exploit this predicted behavior by adopting a pure *run defense*. defending against the run. Randomly passing the ball will keep the Ravens' defense "honest," thereby increasing the likelihood that a *run defense* will be successful. Any systematic pattern of play calls, even if called in the right proportion, will be exploited.

How should the Giants properly randomize its offensive plays? One possibility is to flip a fair coin. Pass the ball if the coin comes up "heads" and run the ball if the coin comes up "tails." Unfortunately, this only works if the optimal mixing rule is 50-50. When applying an optimal mixing rule, players should use a randomizing mechanism that is objective, independent, secret, in the proper proportions, and sufficiently complicated to discourage rivals from imposing patterns on the outcomes.

A modest improvement over flipping a fair coin would be to roll a fair die in which a *pass offense* is called on, say, even numbers and a *run offense* is called on odd numbers. Rolling a fair die also allows for more diverse mixing rules. For example, if the optimal mixing rule is 30-70, the Giants can call a pass play on a roll of 1 or 2 and a run play otherwise. For mixing rules such as 25-75, a pair of dice, which has 21 possible outcomes, could be used. Ignoring the pair 6-6, the Giants could call a pass play for any combination $1-x$, where x goes from 1 to 5, and a run play otherwise.

Of course, we are not likely to see the head coach of the Giants rolling a pair of dice on the sidelines before each play. Much more likely is that each team will put together a randomized game plan prior to kickoff.[2] Alternatively, the Giants' head coach might use another, more discrete, randomizing mechanism, such as calling a pass play when the second hand on his watch is between 1 and 20, and a run play otherwise. Of course, optimal mixing rules should not be a straightjacket. Outside factors, such as field position, defensive alignments, weather, and field conditions should influence play calls, even when it violates the optimal mixing rule.

Even when head coaches in the National Football League know their optimal mixing rules, experience, intuition, and hubris often lead them to dismiss such ideas out of hand. After all, head coaches are hired to manage football games—not to play games of chance. Nevertheless, the logic of randomizing plays to keep the opposition off balance is inescapable.

Bluffing

The lesson to be learned from the above discussion is that it should not matter if a rival knows your optimal mixing rule. There is nothing that a rival can do to exploit this information provided that moves are randomized. Secrecy is essential since knowledge of a rivals' game plan can be exploited. Lies, deceit, and misdirection are essential elements of effective game play. You may know that it is in your rival's best interest to mislead you, but unless you have some way to separate fact from fiction, a player's best game plan is a randomized optimal mixing rule.

Suppose that the Giants signal a *pass offense*, but run the ball. If the Ravens' head coach calls for a pass defense, he has taken the bait. This is an example of a **bluff**, which is an attempt to gain a strategic advantage through misdirection. Although a rival's outward display should not be taken at face value, it may nonetheless convey important strategic information.

An analysis of bluffing involves two important considerations: expected payoffs and unpredictability. Consider, for example, betting strategies in poker. "The poker hand must at all times be concealed behind the mask of inconsistency. The good poker player must avoid set practices and act at random, going so far, on occasion, as to violate the elementary principles of correct play" (McDonald 1950, p. 30). A "tight" player never bluffs, but also seldom wins a large pot because rivals will not "see" (match) or raise his or her bet. A "loose" player who always bluffs will always be "called" on a big bet.

If we have learned anything from our discussion of optimal mixing rules, it is that the best strategy is to randomize bets on good and bad hands. To see why, consider a game of five-card stud-poker between "Wild Bill" Hickok and "Doc" Holiday. Wild Bill plays poker with Doc on a regular basis. Wild Bill has observed that Doc raises 75 percent of the time on a good hand, and calls 25 percent the rest of the time. On the other hand, Doc raises 25 percent of the time on a bad hand, and folds 75 percent of the time. The probabilities of Doc's betting strategies are summarized in Fig. 10.5.

		Hand	
		Good	*Bad*
	Raise	75	25
Bet	*Call*	25	0
	Fold	0	75

Fig. 10.5.

Wild Bill and Doc believe there is a 50 percent chance of being dealt a good or bad hand. Doc's optimal mixing rule depends on the hand he is dealt. Wild Bill can learn something about the probability that Doc has a good hand based on his betting strategy. If Doc folds, Wild Bill knows that Doc has a bad hand. If Doc calls, Wild Bill can be sure that Doc is holding a good hand. In either case, the betting is over and the next hand is dealt. But, if Doc raises his bet, Wild Bill cannot be certain whether Doc has a good or a bad hand. The odds are 3 to 1 that Doc has a good hand. While Doc's decision to raise his bet does not reveal whether he has a good hand or a bad hand, Wild Bill knows more than before the cards were dealt.

The process by which Wild Bill revises his probability estimates that Doc has a good hand is referred to as **Bayesian updating,** which is the process by which players revise their expectations based on a rival's prior moves. Bayesian updating uses **Bayes' theorem** to update the probability of an event after learning that some other event has occurred.

Applying Bayes' theorem, the probability that Doc has a good hand after Wild Bill observes his decision to raise, call, or fold, is equal to the probability that he has a good hand and behaves in a certain way, divided by the probability that he bets in a certain way. For example, if Doc folds, he must have a bad hand. If he calls, his hand must be good. If Doc raises his bet, the probability that he has a good hand is 0.5(75) = 37.5 percent. If Doc raises his bet, the probability that he has a bad hand is 0.5(25) = 12.5 percent. Thus, the total probability of Doc raising his bet is 0.5, or 50 percent. Using Bayes' theorem, the probability that Doc has a good hand based on his decision to raise his bet is equal to 37.5/0.5 = 75 percent. Thus, based on his betting strategy, Wild Bill has revised his probability estimate that Doc has a good hand from 50 percent to 75 percent.

Practice Exercises

10.1 Explain the difference between zero-sum and non-zero-sum games.

10.2 Zero-sum games sometimes have a unique Nash equilibrium in pure strategies. Do you agree with this statement? Explain.

10.3 The matching pennies game is an example of a non-zero-sum game because someone always ends up with a penny. Do you agree? Why not?

10.4 Explain, in your own words, the rationale underlying the minimax theorem.

10.5 Explain the difference between a pure strategy and a mixed strategy.

10.6 Is it possible to have a unique, pure-strategy Nash equilibrium in zero-sum games? In mixed strategies?

10.7 Do you agree that a good way to randomize strategies is to flip a fair coin?

10.8 Is it true that a player who randomizes his or her optimal mixing rule can never be exploited by a rival?

10.9 It is never possible for a player to exploit a rival when knowing his or her optimal mixing rule. Do you agree with this statement? Explain.

10.10 In a zero-sum game, it doesn't matter if a player knows a rival's optimal mixing rule as long as he or she does not have access to the player's randomized game plan. Do you agree with this statement? Why?

10.11 Telephoning a company's customer support is frequently accompanied by a pre-recorded message indicating that the call may be monitored to ensure quality service. Monitoring every call, however, would be prohibitively expensive. What solution would you propose?

10.12 Why does the highway patrol randomize the time and location of speed traps?

10.13 Give an example of bluffing that is not mentioned in the text.

10.14 In poker, explain how a rival's bluff can be used to improve a player's chances of winning.

10.15 Consider the first-down game in Fig. 10.2. Suppose that the probability that the Giants' *run offense* given the Ravens' pass defense improves to 90 percent.

a. If larger payoffs are preferred, what is the optimal mixing rule for both teams?
b. Suppose that each team adopts its optimal mixing rule. What is the expected probability of success for each team?
c. Diagram your results.

10.16 The server in a tennis game can serve to the receiver's forehand or backhand. The receiver must anticipate a forehand or backhand serve. There are four strategy profiles: {*Forehand return*, *Forehand serve*}, {*Forehand return*, *Backhand serve*}, {*Backhand return*, *Forehand serve*}, and {*Backhand return*, *Backhand serve*}. The corresponding probabilities that the receiver will successfully return a serve are (90), (20), (30), and (60). Both individuals adopt their optimal mixing rule.

a. What is the receiver's optimal mixing rule and expected success rate?
b. What is the server's optimal mixing rule and expected success rate?

10.17 Suppose that with practice the receiver in the previous practice exercise improves his ability to return a serve to the backhand side. The corresponding probabilities that the receiver will successfully return a serve are (90), (20), (30), and (70). What affect will this improvement have on your answers to the previous practice exercise?

10.18 What is Scarlett's and Rhett's optimal mixing rules and expected payoffs in the battle-of-the-sexes game in Fig. 5.1?

10.19 Consider the competitive, one-time, static game involving players A and B. There are four strategy profiles: {*A1*, *B1*}, {*A1*, *B2*}, {*A2*, *B1*}, and {*A2*, *B2*}. The respective (A, B) payoffs are (10, 5), (1, 2), (15, 10), and (5, 5).

a. Does this game have a Nash equilibrium in pure strategies?
b. What is the optimal mixing rule for each player?
c. What is the expected payoff for both companies if they adopt their optimal mixing rules?

Notes

1. The Giants' expected payoff from a *pass offense* given a Ravens' *pass defense* (*PD*) is

$$E(\pi_G) = 80 - 0.7PD.$$

The Giants' expected payoff from a *run offense* given a Ravens' *pass defense* is

$$E(\pi_G) = 40 + 0.3PD.$$

Solving, the Ravens' optimal mixing rule is $PD^* = 40$ percent and $RD^* = 1 - PD^* = 60$ percent. Substituting, the Giants' expected success rate of gaining first-down yardage is

$$80 - 0.7(40) = 40 + 0.3(40) = 52\ percent.$$

2. Suppose, for example, that the Giants expect to run at most 20 offensive plays. Using a 30-70 optimal mixing rule, the author generated the sequence R (run), R, R, P (pass), P, R, R, P, R, R, P, P, P, R, P, P, R, R, P, R by rolling a pair of fair dice.

CHAPTER ELEVEN

Continuous Strategies

Introduction

Our analysis of static games has thus far been limited to situations involving a discrete number of pure strategies. In the pricing game in Fig. 2.3, for example, the strategy choices of firms A and B were restricted to just two choices—*high price* or *low price*. Most strategic situations in business, however, involve a potentially infinite number of pure strategies. If charging a *high price* is any price greater than that charged by a rival then there are an infinite number of possible strategies from which to choose. The same applies to levels of output, investment expenditures, spending on research and development, advertising outlays, and so on. In this chapter we will expand our search for a Nash equilibrium to include games involving continuous pure strategies.

Continuous Strategies

Suppose that profit-maximizing firms A and B produce close substitutes. Each firm has a core of loyal customers, but there is also a third group who will purchase from whichever firm charges the lowest price. Both firms have a marginal cost of $0.80 and initially charge a coalition price of $4.20 per unit. To maintain the coalition, each firm adopts the tit-for-tat strategy of charging a price equal to its marginal cost plus 80 percent of the price charged by its rival. Suppose, for example, that firm A were to lower its price to $4.19 to capture the price-sensitive segment of the market. Firm B reacts by lowering its price to $0.80 + 0.8($4.19) = $4.15. Firm A retaliates by cutting its price to $0.80 + 0.8($4.15) = $4.12, and so on until each firm charges $0.80 + 0.8($4.00) = $4.00.

Since tit-for-tat does not rule out future cooperation, this process also works in reverse if the price falls below $4.00. If A were to charge a price of $3.80, for example,

B might respond by charging $0.80 + 0.8($3.80) = $3.84. Firm A could interpret this as a signal of B's desire to return to the *status quo ante* by charging 0.8($3.84) = $3.87, and so on until both firms return to a profit-maximizing competitive price of $4.00. In this example, the Nash equilibrium is {$4.00, $4.00}, which was identified from an infinite number of strategy profiles.

Best-Response (Reaction) Functions

In the pricing game discussed above, each firm's **best-response function** (also known as a *reaction function*) is a formal statement of its profit-maximizing reaction to the price charged by its rival. Eqs. (11.1) and (11.2) are the firms' best-response functions, which are depicted in Fig. 11.1. Because there are theoretically an infinite number of pure strategies, it is not practical to search for dominant strategies. Instead, our search for a Nash equilibrium will proceed by eliminating *dominated* strategies.

$$P_A^*(P_B) = 0.8 + 0.8P_B \qquad (11.1)$$

$$P_B^*(P_A) = 0.8 + 0.8P_A \qquad (11.2)$$

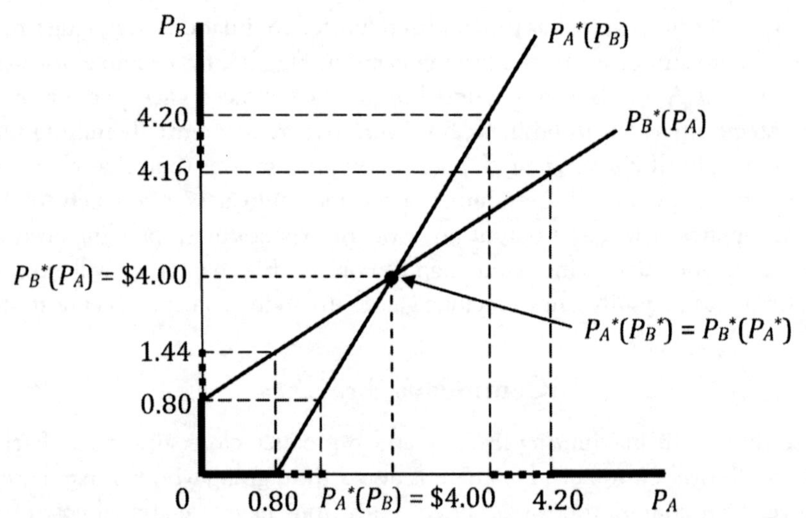

Fig. 11.1.

The search for a Nash equilibrium involves iteratively eliminating dominated strategies. Profit-maximizing firm B in Fig. 11.1 would never have a price less than $0.80 since this would result in an economic loss. Thus, prices less than $0.80 are strictly dominated and are eliminated from further consideration. Similarly, prices above the coalition (monopoly) price of $4.20 are also strictly dominated and should also be eliminated. Thus, B's optimal pricing strategy must lie in the interval [$0.80, $4.20]. The same line of reasoning also applies to firm A since the firms have identical reaction functions. These strictly-dominated strategies are shown as the heavy black line segments along the axes in Fig. 11.1.

Firm B's best response to a price of \$4.20 is \$4.16. Higher prices are strictly dominated and should also be eliminated. Firm B' best response to a price of \$0.80 is \$1.44. Lower prices are strictly dominated and should also be eliminated. These strictly-dominated strategies are shown as the dashed line segments along the axes in Fig. 11.1. After these strictly-dominated strategies are eliminated, B's optimal pricing strategy must now lie in the narrowed interval [\$1.44, \$4.16]. This must also be true of firm A as well. Continuing in this matter, the upper and lower bounds of the interval eventually converge to the intersection of the firms' reaction functions. The Nash equilibrium strategy profile {\$4.00, \$4.00} is more efficiently determined by simultaneously solving Eqs. (11.1) and (11.2).

Although the analysis used to identify a Nash equilibrium strategy profile is sequential, it is important to remember that this is one-time static game in pure strategies. Each firm moves just once. The objective was to identify each firm's best response to a continuum of possible prices charged by its rival.

Solved Exercise

An industry consists of two firms producing an identical product. The market demand for the combined output of the two firms is

$$(Q_1 + Q_2) = 201 - 0.2P. \tag{11.3}$$

Each firm's total cost of production is

$$TC_i = 100 + 5Q_i, \tag{11.4}$$

where i = 1, 2.

a. What is the best-response function for each firm?
b. What is the Nash equilibrium output strategy profile and market price? Calculate each firm's total profit.

Solution

a. Solving Eq. (11.3) for price we obtain

$$P = 1{,}005 - 5(Q_1 + Q_2). \tag{11.5}$$

The total revenue equation for firm 1 is

$$TR_1 = PQ_1 = 1{,}005Q_1 - 5Q_1^2 - 5Q_1Q_2. \tag{11.6}$$

Combining Eqs. (11.4) and (11.6), firm 1's total profit is

$$\pi_1 = TR_1 - TC_1 = -100 + 1{,}000Q_1 - 5Q_1^2 - 5Q_1Q_2. \tag{11.7}$$

The profit-maximizing level of output is found solving the equation

$$\frac{\partial \pi_1}{\partial Q_1} = 1{,}000 - 10Q_1 - 5Q_2 = 0. \tag{11.8}$$

Firm 1's best response function is

$$Q_1^* = 100 - 0.5Q_2. \tag{11.9}$$

Analogously, firm 2's best response function is

$$Q_2^* = 100 - 0.5Q_1. \tag{11.10}$$

b. *Solving Eqs. (11.10) and (11.11), the Nash equilibrium output for each firm is* $Q_1^*(Q_2^*) = Q_2^*(Q_1^*) = 66.67$ *units. Substituting these values into Eq. (11.3), the market price is* $P^* = \$338.39$. *Each firm's profit is* $\pi_1^* = \pi_2^* = \$22{,}121.12$.

Tragedy of the Commons

A well-known application of best-response functions is the "tragedy of the commons," which describes the conflict that arises when individual self-interests clash with the social welfare over the use of jointly-owned resources. Specifically, there is a tendency for individuals to overuse the source, which results in a decline in social welfare. The tragedy of the commons derives its name from a parable by William Forster Lloyd (1833) and popularized by Garrett Hardin (Hardin, 1968; Hardin and Baden, 1977). The tragedy of the commons is at the root of many environmental and government fiscal problems.

To illustrate the problem, suppose that two shepherds graze sheep on the village "commons" (a communal area that is available for use by the public). Since the size of the commons is fixed, the more sheep that are allowed to graze, the greater the competition for the available grass. This is not a problem when the flock is small since the grass growth rate is sufficient to support good nutrition. Initially, the social benefits in terms of wool and mutton production increase with the size of the flock. Beyond some flock size, however, new grass is insufficient to support good nutrition and wool and mutton production declines.

An exclusively owned grazing area will be optimally used because a single shepherd receives all the benefits and incurs all the costs. When multiple shepherds have access to a communal grazing area, however, there is a tendency to overgraze. Suppose, for example, that shepherds A and B form a coalition and graze s_A and s_B sheep, respectively. The social benefits in terms of wool and mutton production is given by the equation

$$B = s(180 - s), \tag{11.11}$$

where $s = s_A + s_B$. The socially optimal number of sheep is $s = 90$, which is found by solving

$$\frac{dB}{ds} = 180 - 2s = 0. \tag{11.12}$$

Substituting this result into Eq. (11.1), the total social welfare is $B = 8{,}100$. Each shepherd will graze 45 sheep and receive half the benefits. These results are depicted in Fig. 11.2.

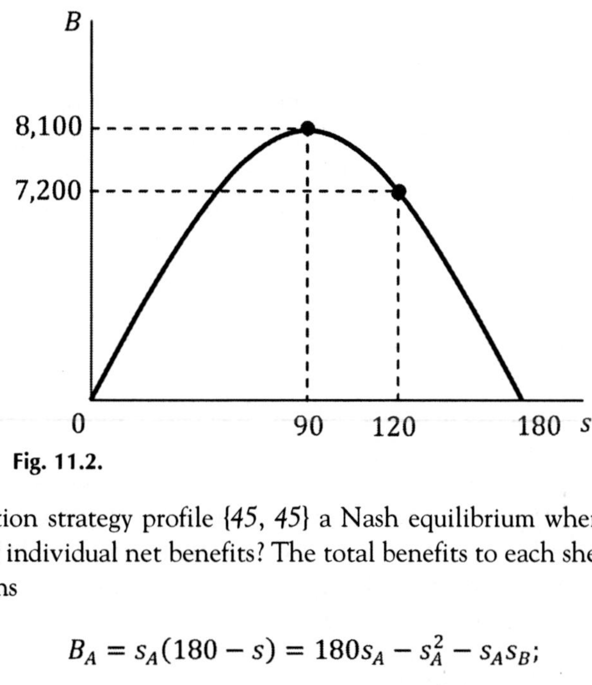

Fig. 11.2.

Is the coalition strategy profile {45, 45} a Nash equilibrium when the shepherds maximize their individual net benefits? The total benefits to each shepherd are given by the equations

$$B_A = s_A(180 - s) = 180s_A - s_A^2 - s_A s_B; \tag{11.13}$$

$$B_B = s_B(180 - s) = 180s_B - s_B^2 - s_A s_B. \tag{11.14}$$

Shepherd A's best-response function is the value of s_A that maximizes total benefits for any value of s_B. Optimizing Eq. (11.14) with respect to s_A we get

$$\frac{\partial B_A}{\partial s_A} = 180 - 2s_A - s_B = 0. \tag{11.15}$$

Solving Eq. (11.15) for s_A, shepherd A's best-response function is

$$s_A^*(s_B) = 90 - \frac{s_B}{2}. \tag{11.16}$$

Similarly, shepherd B's best-response function is

$$s_B^*(s_A) = 90 - \frac{s_A}{2}. \tag{11.17}$$

Solving Eqs. (11.16) and (11.17), the Nash equilibrium is $\{s_A^*(s_B^*), s_B^*(s_A^*)\}$ = {60, 60} resulting in total social benefits of B = 7,200, compared with coalition benefits of B = 8,100. The reason why this game is called a tragedy is because individual initiative does not serve the common good. Only when the shepherds form a coalition and each grazes 45 sheep is the social welfare maximized. Overgrazing occurs because no one owns the communal grazing area. Each shepherd has an incentive to use as much of this scarce resource as possible.

The tragedy of the commons is an example of **market failure**, which occurs when supply and demand fails to generate socially efficient levels of consumption or production. Market failure has been used as a justification for government intervention in the marketplace. This is precisely what occurred in fifteenth- and sixteenth-century England when ownership of common areas was seized by the aristocracy who levied grazing rents to maximize their incomes. Although grazing was reduced to more socially optimal levels, income was redistributed from the peasants to the landed gentry.

A well-known, market-based solution to the tragedy of the commons is to establish well-defined private-property rights (Coase, 1960). Unfortunately, private-property rights are not always easy to define or enforce. It is difficult, for example, to regulate the exploitation of the world's deep-sea resources or the amount of industrial and consumer pollutants spewed into the atmosphere. As a result, whales were indiscriminately hunted to near extinction, and global warming arguably poses a serious threat to the future of mankind. In such cases, an international coalition is required, which is not easy to achieve since this requires concessions that are inconsistent with national political and economic self-interests.

Shifting Best-Response Functions

How will a change in a player's reaction function affect the Nash equilibrium? Suppose that shepherd A in our example develops a more efficient sheep husbandry that reduces costs and increases wool and mutton production. Suppose that shepherd A's revised reaction function is

$$s_A'(s_B) = 120 - \frac{s_B}{2}, \tag{11.18}$$

which is depicted in Fig. 11.3 as a right-shift in shepherd A's reaction function from $s_A^*(s_B)$ to $s_A'(s_B)$. The resulting Nash equilibrium strategy profile is {*100, 40*}. The strategic effect of improved husbandry is a larger flock for shepherd A. Shepherd B's best strategic response is a smaller flock.

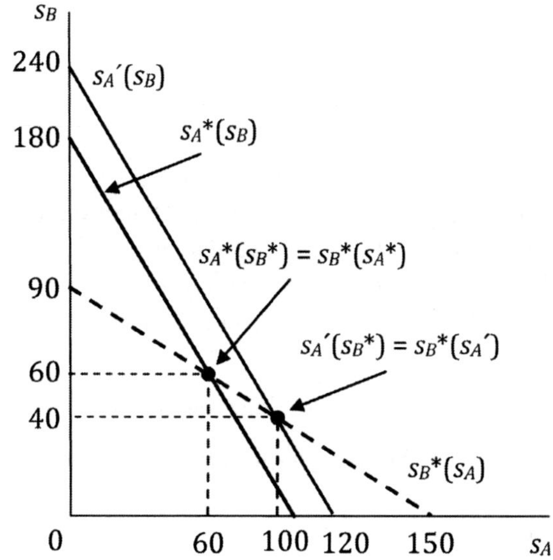

Fig. 11.3.

Linear reaction functions of the type discussed in this chapter suggest that pure strategy, static games involving continuous pure strategies have unique Nash equilibria. There is no guarantee, however, that reaction functions will be linear. Nonlinear reaction functions may result in multiple Nash equilibria.

Solved Exercise

Suppose that market demand is given by Eq. (11.3). Suppose that the total cost equations of two firms in the industry are

$$TC_1 = 100 + 5Q_1; \tag{11.19}$$

$$TC_2 = 100 + 25Q_2. \tag{11.20}$$

a. *What is each firm's best-response function? How do these reaction functions compare with Eqs. (11.9) and (11.10)?*
b. *What is the Nash equilibrium strategy profile, market price, and profit for each firm?*

Solution

a. *Since there is no change in firm 1's total cost, the best-response function remains*

$$Q_1 = 100 - 0.5Q_2. \tag{11.21}$$

By contrast, firm 2's best-response function is

$$Q_2 = 98 - 0.5Q_1, \tag{11.22}$$

which is depicted as a shift in the reaction function to the left.

b. *Solving the firms' reaction functions, the Nash equilibrium strategy profile is {68, 64} compared with {66.67, 66.67} previously. Substituting these new output levels into Eq. (11.5), the market-clearing price is $345, which is predictably higher because industry output is less. Firm 1's profit is greater ($23,080) and firm 2's profit lower ($20,380).*

Practice Exercises

11.1 Explain the difference between static games involving discrete and continuous strategies.

11.2 Define what is meant by a player's best-response function.

11.3 In a static game involving two players with linear reaction functions, how will eliminating dominated strategies lead to a Nash equilibrium?

11.4 In a static game involving two players with continuous strategies, the simultaneous solution of linear reaction functions will always result in a unique Nash equilibrium. Do you agree with this statement?

11.5 What is the tragedy in the tragedy of the commons?

11.6 Why is the tragedy of the commons an example of market failure?

11.7 Many countries experience chronic and severe budget crises associated social welfare programs. French economist Frederic Bastiat (1848) argued that the welfare state institutionalizes "reciprocal plunder." Explain.

11.8 Improved methods of sheep husbandry can shift a shepherd's reaction function.

a. Explain the effects on the benefits of each shepherd.

b. How are the shepherds' benefits affected if they form a coalition?

11.9 Suppose that the benefits from fishing in a lake is

$$B = f(1{,}000 - f), \tag{PE11.9}$$

where f is the number of fish in the lake.

a. What is the maximum number of fish that the lake can support?

b. Suppose that it takes a year for the birth rate of fish to replenish the fishermen's catch. What annual catch will maximize the benefits from fishing? Calculate the total benefits.

c. Suppose that there are two fishermen of equal skill. The annual combined catch is $f = f_1 + f_2$. What is the total annual catch if the two fishermen are rivals? Calculate the total benefits.
d. Suppose that fisherman 1 is twice as skilled at fishing. What is the total annual catch? Calculate total benefits.

11.10 Suppose that the demand for the combined output of two firms is

$$Q_T = 10 - 0.1P, \quad \text{(PE11.10a)}$$

where $Q_T = Q_A + Q_B$. The firms' total costs are

$$TC_A = 4Q_A; \quad \text{(PE11.10b)}$$

$$TC_B = 8Q_B. \quad \text{(PE11.10c)}$$

a. Determine the best-response function for each firm.
b. What is the Nash equilibrium strategy profile for this game? Calculate the market price and profit for each firm.

CHAPTER TWELVE

Static Oligopoly Games

Introduction

The term **oligopoly** refers to an industry that is dominated by a few large firms producing identical or similar products. A special case of an oligopoly is a *duopoly*, which is an industry consisting of two firms. A distinguishing feature of this form of imperfect competition is that a firm's price, output, and other decisions affect, and are affected by, the decisions of rivals.

The pricing and output decisions of oligopolistic firms reflect the nature and degree of competition. To be successful, managers must know as much about their rivals' operations as they do about their own. Moreover, while the entry of new firms into the industry is possible, it is very difficult, which may enable firms to earn positive economic profits in the long run. Joe Bain (1956) indicated that legal, economic, and other barriers represent inherent advantages enjoyed by incumbent firms over latent competitors. George Stigler (1947) argued that these barriers consist of any costs incurred by potential entrants that are not paid by incumbents.

Many barriers to entry are the same as those enjoyed by monopolists, such as ownership of patent and copyright protection. Oligopolists can also restrict entry by controlling distribution outlets and persuading retail chains to carry only its product by offering selective discounts, long-term supply contracts, or gifts to management. Product guarantees also serve as an effective barrier to entry. New car warranties, for example, restrict competition by requiring the exclusive use of authorized parts and service.

Oligopolists may produce perfect or near substitutes. The automobile industry, for example, is comprised of a relatively few large firms producing similar, albeit differentiated, products. Although the Honda Accord and the Ford Taurus are mid-sized sedans, consumers do not view them as identical products. By contrast, some large firms produce a homogenous product. USX and Mitsubishi Steel, for example, are producers of high-quality cold-rolled steel. ExxonMobil and Royal Dutch Shell refine and market gasoline. McDonald's, Wendy's, Carl's Jr., White Castle, Burger King, In-N-Out Burger, Checkers, Sonic, and others sell hamburgers that appeal to different clienteles because their products differ in terms of ingredients, taste, and appearance, and their restaurants differ in style, motif, atmosphere, customer service, packaging, and so on.

The pricing and output decisions of oligopolists affect, and are affected by, the pricing and output decisions of rival firms. This strategic interdependence is illustrated in Fig. 12.1 where *DD* is the market demand for the output of the industry, and *dd* is the demand for the output of an individual firm. If all firms in the industry lower their price from P_1 to P_2, the market quantity demanded increases from Q_1 to Q_2. On the other hand, if only one firm lowers its price from P_1 to P_2, its sales increase from Q_1 to Q_3. This implies that over this price range, the demand for the output of an individual firm is more price elastic than the demand for the output of the industry as a whole. The decision by one firm to unilaterally lower its price to capture a larger market share lies at the heart of most modern analyses of imperfect competition.

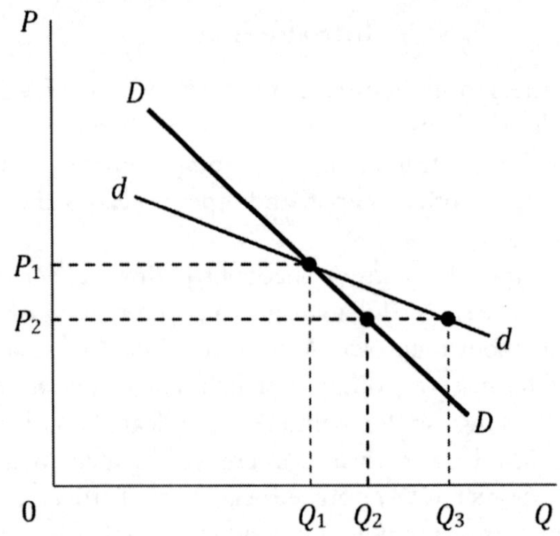

Fig. 12.1.

Cournot Model

In the early-nineteenth century, French mathematician Antoine Augustin Cournot (1838) developed a model to explain how competing firms determined output levels to maximize profits. Although Cournot employed standard optimization techniques, in retrospect his work on the behavior of oligopolies anticipated the concept of a Nash equilibrium by more than a century. In retrospect, Cournot's work represents a seminal contribution to literature on game theory. We will begin our discussion with a graphical description of Cournot's model. We will then reformulate the Cournot model as a static, output-setting game involving continuous strategies.

The **Cournot model** is a static, output-setting game in which rival firms simultaneously determine their optimal output levels. Once determined, output cannot be changed, or at least not at significant economic cost. In the Cournot model, output is the only decision variable. Once the firms' output levels are determined, the market price (P) adjusts to equate consumer demand with total industry supply (Q_T).

The objective of each firm is to maximize profit given the output of rival firms. To simplify the analysis, we will assume that the industry consists of two profit-maximizing firms producing perfect substitutes; the market demand curve is linear. To identify the Nash equilibrium in the Cournot output-setting game it is necessary to specify each firm's best-response function. The Nash equilibrium strategy profile for Cournot output-setting firms occurs at the intersection of these best-response functions. Suppose, for example, that the inverse demand function for the output of the two firms ($Q_T = Q_1 + Q_2$) producing perfect substitutes is

$$P = 25 - 0.5Q_T = 25 - 0.5(Q_1 + Q_2). \quad (12.1)$$

From Eq. (12.1), the firms' total revenue equations are

$$TR_1 = PQ_1 = 25Q_1 - 0.5Q_1^2 - 0.5Q_2Q_1; \quad (12.2)$$

$$TR_2 = PQ_2 = 25Q_2 - 0.5Q_2^2 - 0.5Q_1Q_2. \quad (12.3)$$

The total cost equation for each firm is

$$TC_i = 10Q_i. \quad (12.4)$$

From Eqs. (12.2) and (12.3), the firms' marginal revenue equations are

$$MR_1 = 25 - Q_1 - 0.5Q_2; \quad (12.5)$$

$$MR_2 = 25 - Q_2 - 0.5Q_1. \quad (12.6)$$

From Eq. (12.4), each firm has a marginal cost of $10. To maximize profit, each firm will produce where marginal revenue equals marginal cost. The resulting best-response functions are

$$Q_1^*(Q_2) = 15 - 0.5Q_2; \quad (12.7)$$

$$Q_2^*(Q_1) = 15 - 0.5Q_1. \quad (12.8)$$

Solving Eqs. (12.7) and (12.8), the **Cournot-Nash equilibrium** strategy profile is $\{Q_1^*(Q_2^*), Q_2^*(Q_1^*)\} = \{10, 10\}$. Substituting these output levels into Eq. (12.1), the market-clearing price is P^* = $15. Each firm earns a maximum profit of $\pi_i^* = TR_i - TC_i$ = ($15 × 10) – ($10 × 10) = $50.

Advertising in a Cournot Setting

The Cournot model may be used to understand the strategic significance of a firm's advertising expenditures. Suppose that an industry consists of two output-setting competitors, where spending on advertising (A) by one firm results in increased demand for the output of both firms. Suppose that the market inverse demand for the combined output of both firms is

$$P = 20 + A - 0.5(Q_1 + Q_2). \tag{12.9}$$

Both firms simultaneously and independently choose their respective output levels. Each firm's total cost, which includes its advertising expenditures, is

$$TC_i = 5Q_i + \frac{100}{9}A. \tag{12.10}$$

The total profit equation of firm 1 is

$$\pi_1 = TR_1 - TC_1 = 15Q_1 + 5AQ_1 - 0.5Q_1^2 - 0.5Q_1Q_2 - \frac{100}{9}A. \tag{12.11}$$

The profit-maximizing level of output is found by solving the equation

$$\frac{\partial \pi_1}{\partial Q_1} = 15 + 5A - Q_1 - 0.5Q_2 = 0. \tag{12.12}$$

Firm 1's best-response function is

$$Q_1 = 15 + 5A - 0.5Q_2. \tag{12.13}$$

Similarly, firm 2's best-response function is

$$Q_2 = 15 + 5A - 0.5Q_1. \tag{12.14}$$

Solving Eqs. (12.13) and (12.14) simultaneously, the Cournot-Nash equilibrium is for each firm to produce $Q_1 = Q_2 = 10 + 10A/3$, which depends on the optimal level of advertising expenditures. Substituting these output levels into Eq. (12.11), firm 1's profit equation becomes

$$\pi_1(A) = 50 + \frac{100}{3}A - \frac{50}{9}A^2. \tag{12.15}$$

The optimal level of advertising for firm 1 can be found by optimizing Eq. (12.15), and similarly for firm 2. The profit-maximizing level of advertising for each firm is A = \$3. Substituting these values into the firms' best-response functions, the

Cournot-Nash equilibrium is for each firm to produce $Q_1 = Q_2 = 20$ units of output and earn $\pi_1 = \pi_2 = \$100$.

Bertrand Model

If you are uncomfortable with the output-setting assumption of the Cournot model, you are not alone.[1] While Cournot assumed that firms simultaneously set output levels and price adjusts to clear the market, French mathematician and economist Joseph Bertrand (1883) argued that firms compete directly on the basis of price. The market clears when consumers adjust their purchasing decisions. The **Bertrand model** is a static, price-setting game in which firms cannot switch strategies and industry output adjusts to market demand.

The predictions of the Bertrand model depend on whether the firms produce homogeneous goods (perfect substitutes) or differentiated goods (imperfect substitutes). We will assume that the industry consists of two firms producing perfect substitutes at the same constant marginal cost. We also assume that demand is perfectly elastic in the sense that consumers will always buy from the firm charging the lowest price. Specifically, when $P_1 < P_2$, $Q_1 = Q_1(P_1)$ and $Q_2 = 0$, and *vice versa* for firm 2. We will also assume that if $P_1 = P_2$, $Q_1 = Q_2 = 0.5Q_T$.

What is the optimal pricing strategy for each firm? Suppose initially that the firms form a coalition and charge a monopoly price P^m. Each firm sells to half the market and earns $0.5(P^m - MC)Q_T^m$. In this setting, neither firm has an incentive to charge a higher price. On the other hand, if firm 2 charges a price that is just below P^m and above MC, firm 1's best response is to lower its price just below that of firm 2, sell to the entire market, and earn a positive economic profit. Firm 2 will respond by lowering its price below P_1, and so on until $P = MC$. Firm 1's best response is to charge a price that is just below P_2, but to never charge a price that is above P^m or below MC. Firm 1's best-response function $P_1 = P_1^*(P_2)$ is depicted in Fig. 12.2. For $P_2 > P^m$, $P_1 = P^m$. For $P^m > P_2 > MC$, $P_2 > P_1 > MC$. Finally, for $P_2 < MC$, $P_1 = MC$.

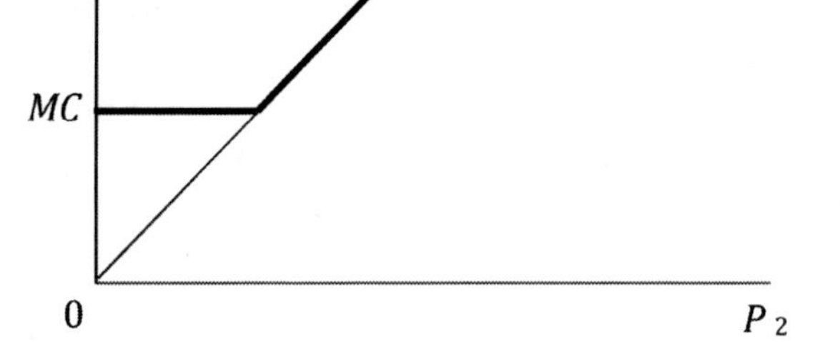

Fig. 12.2.

If $MC_1 = MC_2$, firm 2's best-response function $P_2 = P_2^*(P_1)$ is the same as that for firm 1. These best-response functions are depicted in Fig. 12.3. Since each firm has an incentive to undercut its rival's price, but never charge a price that is less than its marginal cost, the **Bertrand-Nash equilibrium** for this game occurs at point E where $P_1^*(P_2^*) = P_2^*(P_1) = MC$.

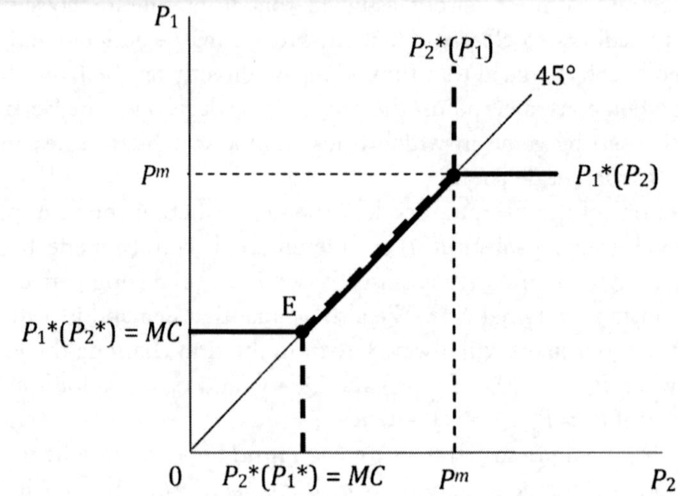

Fig. 12.3.

Bertrand Paradox

The remarkable thing about price competition in industries consisting of two firms producing identical goods and having the same marginal cost is that $P = MC$ and firms earn zero economic profit. This is precisely what we would expect to observe in perfectly competitive industries, which are characterized by a large number of small firms producing an identical product in which entry into, and exit from, the industry is unimpeded. According to the Bertrand model, however, it takes just two large firms producing perfect substitutes to guarantee a perfectly competitive outcome. This prediction is known as the **Bertrand paradox** because it is not observed in the real world. There are three possible explanations for the Bertrand paradox, including *capacity constraints*, *differentiated products*, and *collusion*.

The Bertrand paradox depends crucially on the assumption that competing firms have enough capacity to satisfy total market demand at any price. If this is not the case, each firm will charge a price that is greater than marginal cost, and each will earn a positive economic profit. To see how this explains the paradox, suppose that both firms initially charge $P' > MC$ and split the market. If firm 1 charges a lower price, its sales will double if it has enough capacity to satisfy total market demand, while sales of firm 2 shrink to zero. On the other hand, if firm 1 cannot supply enough output to satisfy total market demand, sales of firm 2 will not fall to zero.

Now, suppose that firm 1 is producing at capacity Q_1^0 when $P' > MC$. Charging $P_1 < P_2 = P'$ does not help since $(P_1 - MC)Q_1^0 < (P' - MC)Q_1^0$. Charging $P_1 > P_2 = P'$ also does not help since profit-maximizing firm 1 is producing where $MR = MC$, which implies that demand is price elastic. Raising price will lower profit because of the decline in total revenues ($MR < 0$) at constant marginal cost. Thus, the optimal price for firm 1 is $P_1 = P'$. By analogous reasoning, $P_2 = P'$ is the optimal price for firm 2. Thus, the Bertrand-Nash equilibrium when both firms are capacity constrained is $P_1 = P_2 = P' > MC$.

The Bertrand paradox can also be explained when the firms produce imperfect substitutes. Two products are close substitutes if at some price a buyer would prefer to purchase from one firm instead of another. In fact, differentiated products come closer to what is actually observed in the real world. Examples of industries in which firms produce differentiated products abound, including clothing, soft drinks, beer, cosmetics, computer hardware, computer software, automobiles, petroleum refineries, fast-food franchises, and so on.

Product differences may be real or imagined. Regular (87-octane) gasoline has a precise chemical composition. Yet, many consumers prefer, say, ExxonMobil or Texaco to other brands. Firms producing differentiated products often commit substantial sums to advertising, promotion, product placement, and packaging to reinforce real and perceived product differences. The goal is to create customer loyalty and reinforce brand-name identification, which increases the firm's market power by making the demand for its product more price inelastic.

One way to model strategic interaction among firms producing differentiated products is to specify separate demand equations for the output of each firm. Suppose, for example, that the demands for the output of duopolists producing differentiated products are

$$Q_1 = 100 - P_1 + 0.5P_2; \quad (12.16)$$

$$Q_2 = 100 - P_2 + 0.5P_1. \quad (12.17)$$

Unlike perfect substitutes, firms producing differentiated products cannot capture the entire market by undercutting its rival's price because some customers have a strong preference for the product of one firm over another. In this setting, a firm that raises its price will experience a decline in the quantity demanded for its product, while the demand for the output of its rival increases as some customers defect to the relatively cheaper imperfect substitute.

From Eqs. (12.16) and (12.17), the firms' total revenue equations are

$$TR_1 = P_1Q_1 = 100P_1 - P_1^2 + 0.5P_1P_2; \quad (12.18)$$

$$TR_2 = P_2Q_2 = 100P_2 - P_2^2 + 0.5P_1P_2. \quad (12.19)$$

Suppose that both firms have a constant marginal cost of \$10. Setting $MR = MC$, the firms' best-response functions are

$$P_1^*(P_2) = 45 + 0.25P_2; \quad (12.20)$$

$$P_2^*(P_1) = 45 + 0.25P_1. \quad (12.21)$$

Solving Eqs. (12.20) and (12.21), the Bertrand-Nash equilibrium strategy profile is $\{P_1^*(P_2^*), P_2^*(P_1^*)\} = \{\$60, \$60\}$. Substituting these results into Eqs. (12.16) and (12.17), each firm produces 70 units and earns economic profits of $\pi_1 = \pi_2 = (\$60 - \$10)70 = \$3{,}500$. Unlike the Bertrand model depicted in Fig. 12.3, price is greater than marginal cost and each firm earns a positive economic profit.

Practice Exercises

12.1 What is a Cournot-Nash equilibrium? Be specific.

12.2 What was Bertrand's criticism of the Cournot output-setting model?

12.3 What is a Bertrand-Nash equilibrium? Be specific.

12.4 According to the Bertrand model, price competition results in each firm earning zero economic profit. What assumptions are necessary for this outcome?

12.5 What is the Bertrand paradox? What modifications can be made to the assumptions of the Bertrand price-setting model to resolve the paradox?

12.6 How does the assumption of differentiated products resolve the Bertrand paradox?

12.7 Explain how collusive price-fixing agreements resolve the Bertrand paradox. Discuss the role of discount rates and the probability that coalitions will end.

12.8 Suppose that the inverse demand equation for an industry's output is

$$P = 55 - Q. \quad (PE12.8)$$

The industry consists of two profit-maximizing firms with constant average total and marginal cost equal to \$5 per unit. Both firms are Cournot competitors.

a. Determine the best-response function for each firm.
b. What is the Cournot-Nash equilibrium?
c. What is the market price and profit for each firm?

12.9 Suppose that an industry consists of two firms that produce a homogeneous product. The market demand and the firms' cost equations are

$$P = 200 - 2(Q_1 + Q_2); \quad (PE12.9a)$$

$$TC_1 = 2Q_1; \quad (PE12.9b)$$

$$TC_2 = 4Q_2. \quad (PE12.9c)$$

a. Calculate each firm's best-response function.
b. What is the Cournot-Nash equilibrium?
c. Calculate the market price and profit for each firm.

12.10 Suppose that the demand equations for the product of two profit-maximizing firms are

$$Q_1 = 50 - 5P_1 + 2.5P_2; \tag{PE12.10a}$$

$$Q_2 = 20 - 2.5P_2 + 5P_1. \tag{PE12.10b}$$

The firms' total cost functions are

$$TC_1 = 2Q_1; \tag{PE12.10c}$$

$$TC_2 = 5Q_2. \tag{PE12.10d}$$

Suppose that these firms are Bertrand competitors.
a. What is each firm's best-response function?
b. What is the Bertrand-Nash equilibrium for this game?
c. What is the output and profit for each firm?

12.11 Suppose that there are two firms producing an identical product. Market demand is

$$Q = 1{,}200 - P. \tag{PE12.11}$$

The marginal cost of product by both firms is MC = 0.
a. Firms 1 and 2 have 200 units and 300 units of capacity, respectively. What is the Bertrand-Nash equilibrium?
b. Given your answer to part a, calculate each firm's total profit.

12.12 The demand functions for the output of Bertrand price-setting competitors are

$$Q_1 = 20 - 2P_1 + P_2; \tag{PE12.12a}$$

$$Q_2 = 20 - 2P_2 + P_1. \tag{PE12.12b}$$

The firms' total cost functions are

$$TC_1 = 10Q_1; \tag{PE12.12c}$$

$$TC_2 = 10Q_2. \tag{PE12.12d}$$

a. What is the Bertrand-Nash equilibrium?
b. What is the output of each firm?
c. What is each firm's profit?

Note

1. The Cournot and Bertrand models have been criticized for the simplicity of their underlying assumptions. As E. H. Chamberlin (1933, p. 46) noted: "When a move by one seller evidently forces the other to make a countermove, he is very stupidly refusing to look further than his nose if he proceeds on the assumption that it will not." Chamberlin realized that mutual interdependence would lead imperfectly competitive firms to form coalitions and price as a monopolist.

CHAPTER THIRTEEN

Strategic Trade Policy

Introduction

In this chapter, continuous strategies are used to analyze the **strategic trade policy**. Prior to the 1980s, the strategic interaction of firms engaged in international trade was largely ignored. The application of game theory to an analysis of such **commercial policy** instruments as import tariffs, import quotas, export subsidies, countervailing duties, domestic content requirements, antidumping laws, and product safety standards was a significant step forward in our understanding of international trade flows.

Discrete Pure Strategies

We begin our analysis of strategic trade policy by considering a pure-strategy static game involving two companies that dominate the global market for a particular product. Suppose aerospace companies A and B are considering investing $10 million to develop a new, technologically advanced, commercial aircraft.[1] Each company has enough production capacity to satisfy global demand, but must capture a significant share of the global market to justify up-front development costs. If only one company enters the market, the present value of its operating profits is $150 million. If both companies enter the market, competition for sales will drive operating profits to zero, in which case both companies lose their initial investment. The game depicted in Fig. 13.1 has two Nash equilibrium strategy profiles: {*Not produce*, *Produce*} and {*Produce*, *Not produce*}.

		B	
		Produce	*Not produce*
A	*Produce*	(−10, −10)	**(150, 0)**
	Not produce	**(0, 150)**	(0, 0)

Profits: (A, B)

Fig. 13.1.

Developing a new aircraft is a risky venture and neither company is willing to commit $10 million without a reasonable expectation that it will cover its development costs. Suppose that the U.S. government gives a $15 million research and development subsidy to A to promote job growth. The revised normal form of this game is depicted in Fig. 13.2.

		B	
		Produce	*Not produce*
A	*Produce*	(5, −10)	**(165, 0)**
	Not produce	(0, **150**)	(0, 0)

Profits: (A, B)

Fig. 13.2.

As a result of the government subsidy, A has a strictly-dominant *produce* strategy and presumably achieves its goal of more jobs. Suppose, however, that the European Union (E.U.) also subsidizes B. This situation is depicted in Fig. 13.3.

		B	
		Produce	*Not produce*
A	*Produce*	**(5, 5)**	(**165**, 0)
	Not produce	(0, **165**)	(0, 0)

Profits: (A, B)

Fig. 13.3.

When each company receives a $15 million subsidy, the Nash equilibrium strategy profile is {*Produce*, *Produce*}. While neither company suffers a loss, U.S. and E.U. taxpayers bear the burden of the subsidy. While taxpayers are worse off, has the global economy been made better off? Air travelers gain because of more efficient transportation. They may also benefit from lower fares due to increased competition. Whether the U.S. and E.U. economies are better off depends on the international distribution of air travelers, and whether the increased benefits exceed the tax burden.

Our analysis of the economic benefits of the subsidy is further complicated as the number of companies vying for sales increases. Fig. 13.4 depicts the no-subsidy case in which the Russian aircraft manufacturer, C, enters the market. In this case, each company loses $15 million by producing a new aircraft. This larger loss results from $10 million in sunk research and development expenditures, and an additional loss of $5 million in operating losses from increased competition and lower prices.

		C			
		Produce		*Not produce*	
		B		**B**	
		Produce	*Not produce*	*Produce*	*Not produce*
A	*Produce*	(−15, −15, −15)	(−10, **0**, −10)	(−10, −10, **0**)	**(150, 0, 0)**
	Not produce	**(0, −10, −10)**	**(0, 0, 150)**	**(0, 150, 0)**	(0, 0, 0)

Profits: (A, B, C)

Fig. 13.4.

This game has multiple Nash equilibrium strategy profiles: {*Not produce*, *Not produce*, *Produce*}, {*Not produce*, *Produce*, *Not produce*}, and {*Produce*, *Not produce*, *Not produce*}. Without government intervention, production of a new aircraft is possible if the companies form a coalition, by forming, say, a joint venture. Alternatively, two companies can agree not to produce in exchange for a share of the producing company's profits.

Suppose that each company receives a $15 million subsidy. This situation is depicted in Fig. 13.5. The Nash equilibrium strategy profile is for all three companies to produce, although each breaks even. These results underscore several important points about the effectiveness of a strategic trade policy.

		C			
		Produce		*Not produce*	
		B		**B**	
		Produce	*Not produce*	*Produce*	*Not produce*
A	*Produce*	**(0, 0, 0)**	(**5**, 0, **5**)	(**5**, **5**, 0)	(**165**, 0, 0)
	Not produce	(0, **5**, **5**)	(0, 0, **165**)	(0, **165**, 0)	(0, 0, 0)

Profits: (A, B, C)

Fig. 13.5.

The use of production subsidies is an effective tool for influencing new product development and the flow of international trade. If the subsidy is sufficiently large, the companies earn greater profits. Workers in the aerospace industry benefit from new job creation. Air travelers benefit from more efficient transportation and lower airfares. Lower transportation costs may also stimulate global commerce and economic growth. Offsetting these benefits is the increased burden on taxpayers who finance the government subsidy.

Continuous Pure Strategies

In the remainder of the chapter we will examine strategic trade policy and continuous pure strategies of profit-maximizing domestic monopolies in the United States and Japan that produce identical products. There is no international trade, perhaps because of prohibitively high trade barriers and transportation costs. To simplify the analysis, we will assume that there are no fixed costs and constant marginal cost is $1 per unit for each company. We will also assume that U.S. and Japanese consumers have identical preferences. The countries' respective inverse demand equations are

$$P_{US} = 5 - \frac{1}{5}Q_{US}^{US}; \quad (13.1)$$

$$P_J = 5 - \frac{1}{5}Q_J^J, \quad (13.2)$$

where P_{US} and P_J are unit prices in the United States and Japan, respectively. The superscript on output denotes the product's origin and the subscript denotes the market in which the output is sold.

The corresponding profit equations are

$$\pi_{US}^{US} = \left(5 - \frac{1}{5}Q_{US}^{US}\right)Q_{US}^{US} - Q_{US}^{US}; \quad (13.3)$$

$$\pi_J^J = \left(5 - \frac{1}{5}Q_J^J\right)Q_J^J - Q_J^J, \quad (13.4)$$

where the subscript represents the company's nationality and the superscript represents the source of its profit. The profit-maximizing level of output for the U.S. company is found by solving

$$\frac{d\pi_{US}^{US}}{dQ_{US}^{US}} = 4 - \frac{2}{5}Q_{US}^{US} = 0. \quad (13.5)$$

The profit-maximizing level of output is $Q_{US}^{US} = 10$ (million) units. From Eq. (13.1), the domestic market-clearing price is $P_{US} = \$3$. From Eq. (13.3), the U.S. company's maximum economic profit is $\pi_{US}^{US} = \$20$ (million). Analogously, $Q_J^J = 10$, $P_J = \$3$, and $\pi_J^J = \$20$.

National Welfare

We will assess the effects of strategic trade policy by examining changes in a country's national welfare, which is defined as the sum of consumer and producer surpluses. **Producer surplus** is a company's operating profit, which is the difference between total revenue and total variable cost. Since there are no fixed costs in the above example, producer surplus equals the firm's total economic profit, which is \$20 million. **Consumer surplus** is the total value that buyers receive from the purchase of a good or service less their total expenditures. Diagrammatically, consumer surplus is the shaded area in Fig. 13.6, which is $0.5[(\$5 - \$3)10] = 0.1(10^2) = \$10$. Thus, pre-trade national welfare in the United States and Japan is $CS + PS = \$10 + \$20 = \$30$ million.

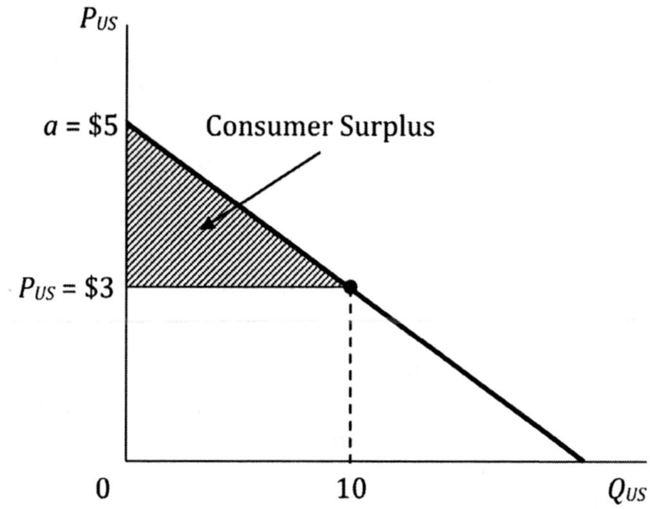

Fig. 13.6.

In general, commercial policies transfer a portion of national welfare between consumers and producers. Assessing the net effect of these transfers can be tricky since it is not possible to say whether a dollar received has the same value as a dollar lost. To overcome the difficulty of evaluating inter-group transfers, we will assume that dollar transfers have equivalent value. Although this one-dollar one-vote metric is imperfect, it allows us to objectively evaluate the effects of commercial policies by comparing changes in the dollar value of consumer and producer surpluses. At least in democratic societies, distributional disputes may in principle be resolved at the ballot box by compelling the "winners" to compensate "losers."

Intraindustry Trade

Neoclassical trade theory, which assumes perfect competition and constant returns to scale, lay at the heart of our understanding of the benefits of international trade.[2] At the core of neoclassical trade theory is the principle of comparative advantage (Ricardo, 1817), which asserts that countries tend to specialize in the production, and export, of goods it can produce at lower opportunity cost. This idea was expanded upon by Eli Heckscher (1919) and Bertil Ohlin (1933), who demonstrated that a country has a comparative advantage in the production of goods that use intensively its relatively abundant factors of production. Later, Wolfgang Stolper and Paul Samuelson (1941) showed that higher prices for exports will positively affect the incomes of owners of factors used intensively in their production and will negatively affect the incomes of owners of factors used less intensively. Together, these ideas provide theoretical justification for the widely-accept notation that countries are made unambiguously better off from an increase in international trade.[3]

Despite the predictions of neoclassical trade theory, however, the data often reveal something quite different. Most international trade is carried out by a relatively small number of highly industrialized countries with similar resource endowments and production technologies. Instead of a movement toward specialization, a significant percentage of cross-border transactions take the form of **intraindustry trade**, which involves the export and import of perfect and imperfect substitutes produced by imperfectly competitive industries that exhibit increasing returns to scale. More importantly, an increase in intraindustry trade may actually lead to a decline in a nation's economic welfare.

Imperfect Competition

Imperfect competition describes a market structure in which the pricing and output decisions of firms affect, and are affected by, the decisions of rivals. To illustrate the effect on national welfare, consider the U.S. and Japanese companies introduced earlier. In the absence of international trade, each company is a monopolist in its home market. With the opening of intraindustry trade, each company must consider output decisions of its rival in the home and foreign markets. The inverse domestic demand equations are now

$$P_{US} = 5 - \frac{1}{5}\left(Q_{US}^{US} + Q_{US}^{J}\right); \qquad (13.6)$$

$$P_{J} = 5 - \frac{1}{5}\left(Q_{J}^{J} + Q_{J}^{US}\right). \qquad (13.7)$$

In addition to its operating expenditures, each company pays \$1 per unit to transport its product to the foreign market. The resulting profit equations, which is the sum of profits earned in the home and foreign markets, are

$$\pi^{US} = \left[5 - \frac{1}{5}(Q_{US}^{US} + Q_{US}^{J})\right] Q_{US}^{US} + \left[5 - \frac{1}{5}(Q_J^J + Q_J^{US})\right] Q_J^{US} - Q_{US}^{US} - 2Q_J^{US}; \quad (13.8)$$

$$\pi^{J} = \left[5 - \frac{1}{5}(Q_J^J + Q_J^{US})\right] Q_J^J + \left[5 - \frac{1}{5}(Q_{US}^{US} + Q_{US}^{J})\right] Q_{US}^{J} - Q_J^J - 2Q_{US}^{J}. \quad (13.9)$$

Eqs. (13.8) and (13.9) will be used to identify how much each company should produce for the home market and how much to export. Consider, for example, the game played in the Japanese market. Maximizing Eqs. (13.8) and (13.9), the companies' best-response functions are

$$Q_J^{US} = 7.5 - \frac{1}{2}Q_J^J; \quad (13.10)$$

$$Q_J^J = 10 - \frac{1}{2}Q_J^{US}. \quad (13.11)$$

Solving Eqs. (13.10) and (13.11), the Cournot-Nash equilibrium is for the U.S. company to produce 3.33 million units and the Japanese company to produce 8.33 million units, which is depicted in Fig. 13.7. As a result of intraindustry trade, total sales have increased from 10 million to 11.66 million units. From Eq. (13.7), the market-clearing price falls from \$3 to \$2.67.

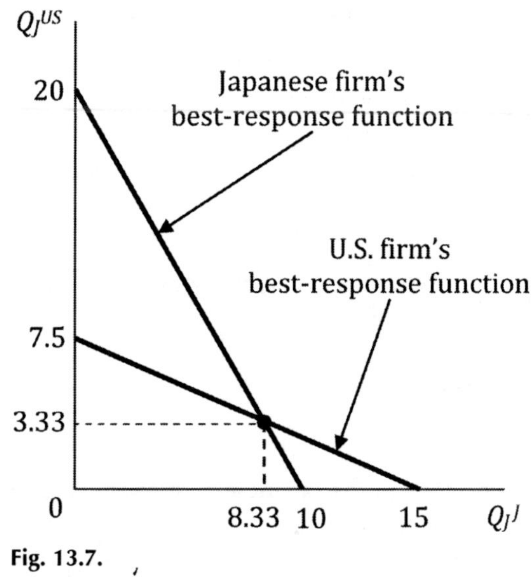

Fig. 13.7.

From Eqs. (13.8) and (13.9), producer surplus with intraindustry trade falls by \$3.88 million to \$16.12 million, of which \$13.89 million is earned in the home market and \$2.23 is earned in the foreign market. This should not be surprising since each company earned monopoly pre-trade profits. Similarly, the Cournot-Nash equilibrium is the same in the U.S. market, but with company output reversed. The lower price and greater quantity demanded results in a \$3.58 increase in consumer surplus to \$13.58 million, which is depicted in Fig. 13.8. The one-dollar one-vote metric suggests that intraindustry trade has made consumers better off, although national welfare falls by \$0.3 million due to the \$3.88 million drop in operating profits.

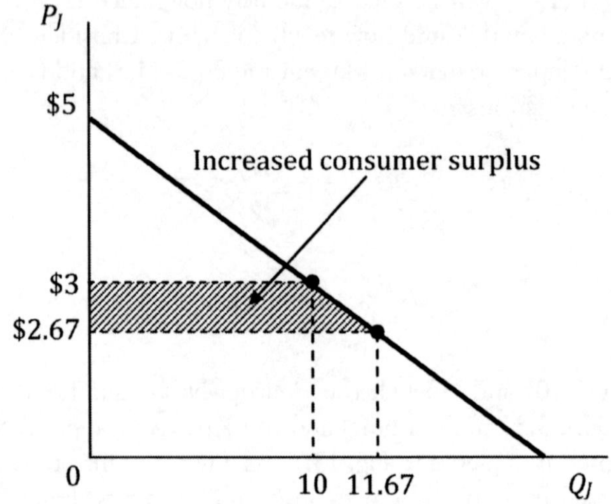

Fig. 13.8.

Intraindustry Coalitions

If producers are made worse off from intraindustry trade then why export? Our earlier discussion of coalition games suggests an explanation. Suppose the U.S. company adopts a *trade* strategy and the Japanese company adopts a *no trade* strategy. What effect will this have on Japanese national welfare? In this case, the U.S. and Japanese profit equations are

$$\pi^{US} = \left(5 - \frac{1}{5}Q_{US}^{US}\right)Q_{US}^{US} + \left[5 - \frac{1}{5}\left(Q_J^J + Q_J^{US}\right)\right]Q_J^{US} - Q_{US}^{US} - 2Q_J^{US}; \tag{13.12}$$

$$\pi^J = \left[5 - \frac{1}{5}\left(Q_J^J + Q_J^{US}\right)\right]Q_J^J - Q_J^J. \tag{13.13}$$

The best-response functions in the Japanese market are given by Eqs. (13.10) and (13.11). Total output in the Japanese market is $Q_J^J + Q_J^{US} = 8.33 + 3.33 = 11.66$ million units. Total output in the U.S. market is $Q_{US}^{US} + Q_{US}^{J} = 10.0 + 0 = 10$ million units. From Eqs. (13.12) and (13.13), operating profits for the {*Trade*, *No trade*} strategy profile are (\$22.22, \$13.89). Analogously, operating profits for the {*No trade*, *Trade*} strategy profile are (\$13.89, \$22.22). These results are summarized in Fig. 13.9. Clearly, domestic companies are made worse off by unilaterally not engaging in intraindustry trade.

		Japan	
		No trade	*Trade*
U.S.	*No trade*	($20, $20)	($13.89, **$22.22**)
	Trade	(**$22.22**, $13.89)	(**$16.12, $16.12**)

Profits: (U.S., Japan)

Fig. 13.9.

The Cournot-Nash equilibrium strategy profile for the coalition game in Fig. 13.9 is {*Trade*, *Trade*}, although both companies are better off by not selling in the foreign market. Forming a coalition to not trade is in the companies' best interest.

What effect does the trade regime have on domestic consumers? According to Fig. 13.10, consumers in both countries are unambiguously better off from intraindustry trade. In this case, however, it is not in the best interest of consumers to form a coalition to achieve some other strategy profile.

		Japan	
		No trade	*Trade*
U.S.	*No trade*	($10, $10)	($13.58, **$10**)
	Trade	(**$10**, $13.58)	(**$13.58, $13.58**)

Consumer surplus: (U.S., Japan)

Fig. 13.10.

The effect on national welfare from alternative trade regimes is summarized in Fig. 13.11, which combines Figs. 13.9 and 13.10. Using the one-dollar one-vote metric, intraindustry trade makes both countries worse off.

		Japan	
		No trade	*Trade*
U.S.	*No trade*	($30, $30)	($27.47, **$32.22**)
	Trade	(**$32.22**, $27.47)	(**$29.70, $29.70**)

National welfare: (U.S., Japan)

Fig. 13.11.

It is almost axiomatic that countries benefit from international trade. This is not the case in our example since output decisions are made by monopolists in the no-trade scenario and Cournot output-setting duopolists that are engaged in intraindustry trade. The companies depicted in Fig. 13.9 are caught in a prisoner's dilemma in which the Nash equilibrium strategy profile {*Trade*, *Trade*} results in lower operating profits for both companies. This suggests that firms have an incentive to seek government assistance, such as export subsidies, which is borne by consumers in the form of higher taxes.

Export Subsidies

An **export subsidy** is a government payment to a domestic company to boost overseas sales of a domestically produced good. Export subsidies can take a variety of forms, such as low-interest loans, lower input prices, and tax relief. Suppose initially that the U.S. government provides a \$0.5 per unit subsidy for exports to Japan, but the Japanese government does not subsidize exports to the United States. As before, transportation costs are \$2 per unit. Eq. (13.14) is now the profit equation of the U.S. company in the Japanese market.

$$\pi_J^{US} = \left[5 - \frac{1}{5}\left(Q_J^{US} + Q_J^J\right)\right] Q_J^{US} - 2Q_J^{US} + 0.5Q_J^{US}. \tag{13.14}$$

The size of this subsidy was chosen to maximize exporter profits in the foreign market. To see how this was determined, this company's profit equation with a per-unit export subsidy (S_{US}) may be written

$$\pi_J^{US} = \left[5 - \frac{1}{5}\left(Q_J^{US} + Q_J^J\right)\right] Q_J^{US} - 2Q_J^{US} + S_{US}Q_J^{US}. \tag{13.15}$$

The U.S. company's best-response function in the Japanese market is

$$Q_J^{US} = 7.5 - \frac{1}{2}Q_J^J + \frac{5}{2}S_{US}. \tag{13.16}$$

The best-response of the Japanese company in its home market is given by Eq. (13.11). Solving these best-response functions we get

$$Q_J^{US} = \frac{10}{3} + \frac{10}{3}S_{US}; \tag{13.17}$$

$$Q_J^J = \frac{25}{3} - \frac{5}{3}S_{US}. \tag{13.18}$$

Substituting Eqs. (13.17) and (13.18) into

$$\pi_J^{US} = \left[5 - \frac{1}{5}\left(Q_J^J + Q_J^{US}\right)\right] Q_J^{US} - 2Q_J^{US} \tag{13.19}$$

we obtain

$$\pi_J^{US} = \frac{20}{9} + \frac{10}{9} S_{US} - \frac{10}{9} S_{US}^2. \tag{13.20}$$

Maximizing Eq. (13.20), the profit-maximizing subsidy is S_{US}, = \$0.50. Total profit of the Japanese company is given by Eq. (13.9). The total profit of the U.S. company is

$$\pi^{US} = \left[5 - \frac{1}{5}\left(Q_{US}^{US} + Q_{US}^{J}\right)\right] Q_{US}^{US} + \left[5 - \frac{1}{5}\left(Q_J^J + Q_J^{US}\right)\right] Q_J^{US} - Q_{US}^{US} - 1.5 Q_J^{US}. \tag{13.21}$$

The companies' best-response functions are

$$Q_J^{US} = 8.75 - \frac{1}{2} Q_J^J; \tag{13.22}$$

$$Q_J^J = 10 - \frac{1}{2} Q_J^{US}. \tag{13.23}$$

The Cournot-Nash equilibrium for this game is for the U.S. company to sell 5 million units and for the Japanese company to sell 7.5 million units in the Japanese market. The market-clearing price is \$2.5. U.S. and Japanese company profits in the Japanese market are \$5 million and \$11.25 million, respectively. Total U.S. company profit with the subsidy increases from \$16.11 million to \$18.89 million, while total profit of the Japanese company falls to \$13.47 million.

How are consumers in both countries affected by the U.S. subsidy? Recall that domestic consumers are taxed to pay for the subsidy. To simplify the analysis, assume that the decline in disposable income has no effect on demand and that consumer surplus is reduced by the amount of the tax. The lower price and increased sales increases consumer surplus in the Japanese market by \$2.04 million to \$15.63 million. Consumer surplus in the United States falls to $CS_{US} - S_{US}$ = \$11.09 million. With the subsidy, U.S. national welfare increases by \$0.28 million to \$29.98 million. National welfare in Japan declines by \$0.6 million to \$29.10 million.

Reciprocity

Suppose the Japanese company petitions its government for an optimal export subsidy. The resulting profit equation is

$$\pi^J = \left[5 - \frac{1}{5}\left(Q_J^J + Q_J^{US}\right)\right] Q_J^J + \left[5 - \frac{1}{5}\left(Q_{US}^{US} + Q_{US}^{J}\right)\right] Q_{US}^J - Q_J^J - 1.5 Q_{US}^J. \tag{13.24}$$

From Eqs. (13.21) and (13.24), the best-response functions in the U.S. market are

$$Q_{US}^{J} = 8.75 - \frac{1}{2}Q_{US}^{US}; \tag{13.25}$$

$$Q_{US}^{US} = 10 - \frac{1}{2}Q_{US}^{US}. \tag{13.26}$$

The decision to subsidize exports depends on what each government is trying to accomplish. From Eqs. (13.25) and (13.26), the profit-maximizing levels of output in the U.S. market are $Q_{US}{}^{US} = 7.5$ and $Q_{US}{}^{J} = 5$. From Eqs. (13.21) and (13.24), each company earns a subsidized profit of \$16.25 million, compared to \$16.12 million without the subsidy. These results are summarized in Fig. 13.12. If the goal of commercial policymakers is to enrich producers, the Cournot-Nash equilibrium for this game is {*Subsidy*, *Subsidy*}.

		Japan	
		No Subsidy	*Subsidy*
U.S.	*No subsidy*	(\$16.12, \$16.12)	(\$13.47, **\$18.89**)
	Subsidy	(**\$18.89**, \$13.47)	(**\$16.25, \$16.25**)

Profits: (U.S., Japan)

Fig. 13.12.

What effect does the subsidy have on consumer welfare? Fig. 13.13 summarizes consumer surpluses for each export subsidy strategy profile. When taxpayers in both countries finance the export subsidy, consumer surplus is \$15.63 – \$2.5 = \$13.13 million, compared with \$13.58 million without the export subsidy. If the objective of commercial policymakers is to make consumers better off, the Cournot-Nash equilibrium strategy is {*No subsidy*, *No subsidy*}.

		Japan	
		No Subsidy	*Subsidy*
U.S.	*No subsidy*	(**\$13.58, \$13.58**)	(**\$15.63**, \$11.09)
	Subsidy	(\$11.09, **\$15.63**)	(\$13.13, \$13.13)

Consumer surplus: (U.S., Japan)

Fig. 13.13.

Finally, Fig. 13.14 summarizes each country's national welfare from alternative export subsidy profiles. If the goal of commercial policymakers is to improve national welfare, each country is caught in a complicated prisoner's dilemma. From Fig. 13.11, the Cournot-Nash equilibrium is for both countries to engage in intraindustry trade, although they are better off by not trading. From Fig. 13.14, the Cournot-Nash equilibrium is subsidized intraindustry trade, even though both countries are better off by not subsidizing exports.

		Japan	
		No Subsidy	*Subsidy*
U.S.	*No subsidy*	($29.70, $29.70)	($29.10,**29.98**)
	Subsidy	(**29.98**, $20.10)	**($29.38, $29.38)**

National welfare: (U.S., Japan)

Fig. 13.14.

The analysis presented in this section is by no means general. The effects of an export subsidy on intraindustry trade depend on the country's respective demand conditions, underlying production technologies, transportation costs, and so on. Moreover, our interpretation of the numerical results depends on the confidence we have in the one-dollar one-vote metric. In spite of this, the above analysis underscores the political difficulties associated with formulating an effective commercial policy in international markets characterized by imperfect competition.

Since the end of World War II there has been a systematic reduction or elimination of trade barriers such as import tariffs, import quotas, and export subsidies. Because of the prisoner's dilemma, freer trade from bilateral trade negotiations would have been problematic since the best interests of producers with deep pockets routinely trump the best interests of many small consumers with little political clout. The systematic reduction of trade barriers was made possible by multilateral negotiations under the auspices of the General Agreement on Tariffs and Trade (GATT), and later the World Trade Organization (WTO). Today, export subsidies are less than a tenth of what they were in the years leading up to World War II.

Practice Exercises

13.1 How do the predictions of an analysis of strategic trade policy and neoclassical trade theory differ, and why?

13.2 How is the one-dollar, one-vote metric used to evaluate national welfare from changes in international trade policy? What are its strengths and weaknesses?

13.3 It is generally thought that free international trade is preferred to autarky. Does the analysis of intraindustry trade presented in this chapter support this belief?

13.4 Which group is most likely to benefit from government intervention to assist domestic producers engaged in international trade? Will the country as a whole be better off?

13.5 In your view, what should be the objective of strategic trade policy? Do you believe that this is, in fact, the objective of trade policy in reality? Why?

13.6 Suppose that intraindustry trade between the United States and Japan involves just two companies. The inverse demand equations for an identical product are

$$P_{US} = 9 - \frac{1}{3}Q_{US}; \qquad \text{(PE13.6a)}$$

$$P_J = 9 - \frac{1}{3}Q_J, \qquad \text{(PE13.6b)}$$

where $Q_{US} = Q_{US}^{\ US} + Q_{US}^{\ J}$ and $Q_J = Q_J^{\ J} + Q_J^{\ US}$. Neither company incurs fixed cost, marginal cost is $1 per unit, and transportation and insurance is $3 per unit. Sales are in millions of units.

a. What is the Cournot-Nash equilibrium in each market?
b. Compare producer and consumer surpluses before and after intraindustry trade. Who are the winners and losers? Are these countries better off as a result of intraindustry trade?
c. Suppose that Japanese trade officials subsidize exports of the Japanese company. What is the optimal per unit subsidy?
d. Suppose that the Japanese government pays an export subsidy, but the U.S. government does not. Who are the winners and losers?
e. Who are the winners and losers if the United States retaliates with its own export subsidy?
f. What is the Cournot-Nash equilibrium if the objective of commercial policy is to improve national welfare?

13.7 Suppose that intraindustry trade between countries A and B involves just two companies. The inverse demand equations for an identical product in the two countries are:

$$P_A = 25 - \frac{1}{5}Q_A; \qquad \text{(PE13.7a)}$$

$$P_B = 10 - \frac{1}{2}Q_B. \qquad \text{(PE13.7b)}$$

where $Q_A = Q_A^A + Q_A^B$ and $Q_B = Q_B^B + Q_B^A$. Neither firm incurs fixed cost, marginal cost is \$2 per unit, and transportation and insurance is \$1 per unit.

a. What is the Cournot-Nash equilibrium?
b. How much will each company earn from intraindustry trade?
c. If the objective of country B is to maximize company B profits, what is the optimal export subsidy? What is the optimal export subsidy for company A?
d. Should the government subsidize exports if the objective is to maximize national welfare?

Notes

1. This game was inspired by the real-world competition between the world's largest producers of commercial aircraft—the Boeing Company of the United States and Airbus S.A.S, which is owned by a consortium of European stakeholders from Great Britain, France, Germany, and Spain.

2. Neoclassical trade theory assumes perfect competition, which means that firms engaged in international trade are "price takers." The constant-returns-to-scale assumption rules out the emergence of a natural monopoly, which means that a firm with enough capacity will be able to satisfy total market demand at lowest per-unit cost.

3. *Autarky* is a national policy of self-reliance in international trade. This term will be used in the present context to describe the no-trade scenario.

CHAPTER FOURTEEN

Product Differentiation

Introduction

Product differentiation is a strategy that can shed light on the decision-making process in many business situations. Managers of firms in monopolistically competitive industries attempt to gain a competitive advantage by selling products that are somewhat different from the products sold by rival firms. Product differentiation may take many forms, including functionality, substance, appearance, packaging, customer service, and location. By differentiating its product, a relatively small firm may be able to exercise control over its prices by carving out a market niche that caters to the preferences of a subset of consumers.

Horizontal Differentiation

Firms can differentiate their products horizontally and vertically. **Horizontal differentiation** refers to a change in a firm's product that makes it better in the eyes of some consumers, but worse in the eyes of others. Arm & Hammer, for example, horizontally differentiates its toothpaste by adding peroxide and baking soda for extra whitening. While individual buyers may disagree about the value of this enhancement, some consumers are willing to pay a higher price for a brighter smile.

Vertical Differentiation

Vertical differentiation refers to changes in a firm's product that makes it better in the eyes of all consumers. A new computer with a free service warranty is better for everyone, for example, than when the service warranty must be purchased. One way for a firm to vertically differentiate its product is by locating close to its customers. This is a recognizable improvement for everyone since it reduces buyers' search and travel costs.

Location

The effect of proximity on sales and revenues can be explained with the help of Fig. 14.1, which depicts two mobile hot-dog vendors' concessions (*Nathan* and *Kayem*), which are initially located at opposite ends of a 5-mile, east-west stretch of beach. *Nathan* is located at the west end and *Kayem* is located at the east end of the beach.

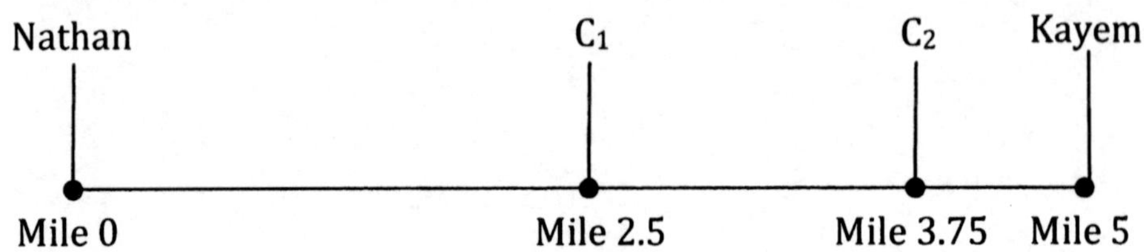

Fig. 14.1.

Other things being equal, hot-dog buyers consider two factors when deciding which concession to patronize—price and location. Assume that both vendors charge \$2 per hot-dog, and that customer travel cost is \$0.10 per mile. We will also assume that hot-dog buyers are uniformly distributed along the beach.

Because travel is costly, we can predict that *Nathan* and *Kayem* will split the market. Hot-dog buyers to the west of mile 2.5 will patronize *Nathan*, and hot-dog buyers to the east of mile 2.5 will patronize *Kayem*. Since there is no difference in price, customer C_1 is indifferent between buying from *Nathan* or *Kayem*. In this example, the only factor that differentiates hot-dogs in the eyes of hot-dog buyers is location.

Suppose that *Nathan* lowers its price to \$1.75, while *Kayem* continues to charge \$2. What effect will this have on sales? Consider a hot-dog buyer located m miles from *Nathan* and $5 - m$ miles from *Kayem*. The cost of a hot-dog from *Nathan* is $\$1.75 + \$0.10m$ and the cost of a hot-dog from *Kayem* is $\$2.00 + \$0.10(5 - m)$. Customer C_2, who is located at mile 3.75 is indifferent between buying from *Nathan* or *Kayem* since $\$1.50 + \$0.10(3.75) = \$2.00 + \$0.10(5 - 3.75) = \$2.125$. *Nathan* will sell hot-dogs to buyers located to the west of mile 3.75 and *Kayem* will sell hot-dogs to buyers to the east of mile 3.75.

But, why should *Nathan* charge \$1.75 and attract customers to the west of mile 3.75 when it can attract buyers from the entire beach by charging a price of \$1.50? If *Nathan* and *Kayem* are Bertrand price-setting competitors, *Kayem* will also respond by charging \$1.50. The result is that *Nathan* and *Kayem* return to the *status quo ante* by splitting the market, but earning less profit. If per unit operating cost is \$0.50, *Nathan* and *Kayem* will each earn \$1.00 from each hot-dog sold.

Since there is nothing to be gained by lowering price, what else can they do to increase sales? One possibility is to relocate. Suppose in Fig. 14.1 that 500 hot-dog buyers are uniformly distributed along the beach. *Nathan* reasons that it can increase its sales by relocating a little further east. In this way, *Nathan* can expect to sell to all buyers to the west of its location and half the buyers between it and *Kayem*. The best response by *Kayem* is to move a little further west and sell hot-dogs to all buyers to the east of its location and half the buyers between it and *Nathan*. *Nathan* will counter by moving still further east, and so on. By iteratively eliminating strictly-dominated location strategies, the Nash equilibrium strategy profile is for *Nathan* and *Kayem* to locate at mile 2.5.

The above analysis can be easily extended to more than two hot-dog vendors. Suppose in Fig. 14.1 that there are three hot-dog vendors located at mile 0, mile 2.5, and mile 5. Since hot-dog buyers are uniformly distributed along the beach, each vendor sells to one-third the market. As before, however, the vendors located at miles 0 and 5 will attempt to increase sales by moving toward the center, until all three vendors are located at mile 2.5.

A variation of this one-time, static game in continuous strategies is depicted in Fig. 14.2. In this game, the beach is divided into five one-mile segments. Each vendor must decide in which mile segment to locate. Since hot-dog buyers are uniformly distributed, each mile segment has 100 customers. The vendors' operating profit depends on their relative locations. Suppose, for example, that *Nathan* locates in mile segment 3 and *Kayem* locates in mile segment 5. *Nathan* will sell 300 hot-dogs to customers in mile segments 1-3, and 50 hot-dogs to customers in mile segment 4, for a total operating profit of $350. *Kayem* will sell 100 hot-dogs to customers in mile segment 5, and 50 hot-dogs to customers in mile segment 4, for a total operating profit of $150. *Nathan* and *Kayem* will split the market and earn $250 by locating in the same mile segment.

		Kayem				
		1	*2*	*3*	*4*	*5*
	1	(250, 250)	(100, **400**)	(150, 350)	(200, 300)	(250, 250)
	2	(**400**, 100)	(250, 250)	(200, **300**)	(250, 250)	(300, 200)
Nathan	*3*	(350, 150)	(**300**, 200)	**(250, 250)**	(**300**, 200)	(350, 150)
	4	(300, 200)	(250, 250)	(200, **300**)	(250, 250)	(**400**, 100)
	5	(250, 250)	(200, 300)	(150, 350)	(100, **400**)	(250, 250)

Operating profits: (*Nathan*, *Kayem*)

Fig. 14.2.

The Nash equilibrium strategy profile for this one-time static game in discrete pure strategies is {3, 3}. By eliminating dominated strategies, *Nathan* and *Kayem* will locate in mile segment 3, which is in the center of the 5-mile stretch of beach. In fact, by eliminating strictly-dominated continuous strategies, *Nathan* and *Kayem* will locate at mile 2.5.

The location games in Figs. 14.1 and 14.2 are an application of the *minimax theorem* (von Neumann, 1928), which says that players in a zero-sum games simultaneously maximize their gains and minimize their losses. These games are zero-sum since *Nathan* and *Kayem* are competing over the distribution of $500. A gain in operating profit by one vendor is an equivalent loss by the other. The Nash equilibrium in these games is for each vendor to locate at the center of the beach. This analysis helps explain why fast-food restaurants, gasoline stations, and similar retail outlets tend to cluster around a centralized location, such as shopping malls. This strategy appears counterintuitive since they are vying for the same group of customers.

Product Differences

The above analysis of product differentiation by location can be easily extended to actual product differences, such as can be found in the cosmetics, soft drinks, soap, and snack foods industries. Other things being equal, including price, firms will attempt to maximize their gains and minimize their losses by differentiating their products. Suppose, for example, that two snack companies differentiated their potato chips by adding barbecue sauce, honey mustard, sea salt, onion dip, and so on until each has 50 percent of the market. If consumer tastes are randomly and uniformly distributed, any attempt by one firm to capture a larger market share by further differentiation will be met with a similar countermove by its rival to reestablish its position in the market. New entrants into this market can only succeed by further differentiating their potato chips to peel away customers from incumbent snack companies. In the absence of barriers to entry, the predicted Nash equilibrium is for the firms to have equal market shares.

Practice Exercises

14.1 Suppose that two soda vendors are located at opposite ends of a 10 mile beach. Vendor A is located at the west end (mile 0) and vendor B is located at the east end (mile 5). Travel cost for soda buyers, who are uniformly distributed along the beach, is $0.15 per mile.

a. Suppose that each vendor sells sodas for $0.75. At what mile will soda buyers be indifferent between buying from either vendor?

b. Suppose that vendor B lowers the price of soda to $0.65. At what mile will soda buyers be indifferent between buying from either vendor?

14.2 Suppose that two hot-dog buyers are not uniformly distributed along a 10-mile beach.

a. Where should vendors A and B locate if 50 percent of hot-dog buyers are located to the west of mile 1 and 50 percent are located to the east of mile 1?

b. Suppose that x hot-dog buyers are located to the west of mile 1 and 45 beach goers who buy hot-dogs are uniformly distributed to the east of mile 1. For what values of x will A and B locate between mile 0 and mile 1?

c. Suppose 5 hot-dog buyers are located to the west of mile 1 and 45 hot-dog buyers are uniformly distributed to the east of mile 2. For what values of x will A and B locate between miles 1 and 2?
d. Suppose 35 hot-dog buyers are located to the west of mile 2 and 45 hot-dog buyers are uniformly distributed to the east of mile 3. For what values of x will A and B locate between miles 2 and 3?

CHAPTER FIFTEEN

Strategic Complements

Introduction

It is often the case in business that a firm's success depends not only on the contributions of individual decisions makers, but on how well decision makers work together. Players are described as **strategic complements** when teamwork yields positive benefits (Bulow et al., 1985). More precisely, suppose that a firm's total profit, which depends on the efforts of two partners, is

$$\pi = \pi(x_1, x_2). \tag{15.1}$$

The firm's profit depends not only on the positive marginal contribution of each partner

$$\pi_1 = \frac{\partial \pi}{\partial x_1} > 0;\ \pi_2 = \frac{\partial \pi}{\partial x_2} > 0, \tag{15.2}$$

but also on how well the partners work with each other

$$\frac{\partial \pi_1}{\partial x_2} > 0;\ \frac{\partial \pi_2}{\partial x_1} > 0. \tag{15.3}$$

An interesting application of strategic complements involves the *principal-agent problem*, which occurs when one group of stakeholders (agents) fails to represent the best interests of another group of stakeholders (principals) because their interests are not properly aligned. Many modern large corporations, for example, are characterized by absentee ownership. Managers are hired to run the company in the owner's (shareholders) best interest, which is generally taken to mean maximizing the value of the firm. Managers, in turn, hire workers to produce goods and services for sale in the market. Since hard work and leisure are substitutes, compensation in the form of fixed salaries and wages will not provide managers and workers with a sufficient incentive to put forth their best efforts on behalf of owners and managers.

To overcome the principal-agent problem, it is necessary to link a stakeholder's compensation to the company's performance with, say, profit-sharing, stock options, and performance bonuses. Although such incentive schemes boost company performance and shareholder returns, they do not necessarily maximize a firm's value. This is because individual self-interested behavior may not be entirely eliminated. When stakeholders are strategic complements, the absence of teamwork may actually result in a less-than-optimal outcome. The reason for this is double marginalization.

Double Marginalization

To illustrate the potential problems that can arise when the welfare of the group depends on the mutually reinforcing behavior by individual members, suppose that a firm produces two goods that are demand complements. An increase in the sales of one good (x_1) is accompanied by an increase in the sales of the other good (x_2), such as skis and ski boots, kites and kite string, and computer hardware and software. Suppose that this firm is organized into two divisions which operate as separate profit centers. The divisions' profit equations are

$$\pi_1(x_1, x_2) = 2x_1 + cx_1x_2 - 0.5x_1^2; \tag{15.4}$$

$$\pi_2(x_1, x_2) = 2x_2 + cx_1x_2 - 0.5x_2^2. \tag{15.5}$$

The firm's total profit is the sum of the divisions' profits, which is

$$\pi(x_1, x_2) = \pi_1 + \pi_2 = 2x_1 + 2x_2 + 2cx_1x_2 - 0.5x_1^2 - 0.5x_2^2. \tag{15.6}$$

The term $2cx_1x_2$ represents to joint contribution of related sales to the firm's total profit. The marginal contributions of each division's output to the firm's total profit are

$$\frac{\partial \pi}{\partial x_1} = \pi_1'(x_1, x_2) = 2 + 2cx_2 - x_1; \tag{15.7}$$

$$\frac{\partial \pi}{\partial x_2} = \pi_2'(x_1, x_2) = 2 + 2cx_1 - x_2. \tag{15.8}$$

From Eqs. (15.7) and (15.8) we obtain

$$\frac{\partial \pi_1'}{\partial x_2} = \frac{\partial \pi_2'}{\partial x_1} = 2c. \tag{15.9}$$

These two goods are strategic complements whenever $c > 0$, which measures the divisions' joint contribution to the firm's total profit. The problem of **double marginalization** occurs when quasi-independent upstream and downstream divisions of the same firm set the price of their output above marginal cost to maximize their respective profits at the expense of maximizing company profits. This is because higher marginal cost of the input produced by the upstream division is passed along to the downstream division that produces the final good.

Given the degree of complementarity (c), how much should each division produce if the objective is to maximize their individual profit? Maximizing Eqs. (15.4) and (15.5), the divisions' best-response functions are

$$x_1^*(x_2) = 2 + cx_2; \tag{15.10}$$

$$x_2^*(x_1) = 2 + cx_1. \tag{15.11}$$

Solving Eqs. (15.10) and (15.11), the equilibrium strategy profile for this game is

$$\{x_1^*(x_2^*), x_1^*(x_2^*)\} = \left\{\left(\frac{2}{1-c}\right), \left(\frac{2}{1-c}\right)\right\}. \tag{15.12}$$

Substituting the equilibrium values into Eq. (15.6) we get

$$\pi^* = \frac{4}{(1-c)^2}. \tag{15.13}$$

How does this compare when the objective is to maximize the firm's total profit? From Eqs. (15.7) and (15.8), the divisions' best-response functions are

$$\hat{x}_1(x_2) = 2 + 2cx_2; \tag{15.14}$$

$$\hat{x}_2(x_1) = 2 + 2cx_1. \tag{15.15}$$

Solving Eqs. (15.14) and (15.15), the equilibrium strategy profile is

$$\{\hat{x}_1(\hat{x}_2), \hat{x}_2(\hat{x}_1)\} = \left\{\left(\frac{2}{1-2c}\right), \left(\frac{2}{1-2c}\right)\right\}. \tag{15.16}$$

At these equilibrium values, the firm's maximum joint profit is

$$\hat{\pi} = \frac{4}{1-2c}. \tag{15.17}$$

Profit from Eq. (15.17) is greater than profit from Eq. (15.13) whenever the degree of complementarity is in the interval $0 < c < 0.5$ (why?). The reader should verify that

$$\frac{2}{1-2c} > \frac{2}{1-c}. \tag{15.18}$$

Condition (15.18) says that each division will produce more, and earn greater profit, when the objective is to maximize the firm's profit, rather than to organize the company into quasi-independent profit centers. According to these results, double marginalization distorts incentives that result in a less-than-optimal outcome.

Starting from our original assumption of double marginalization, increasing output in the above example is beneficial, but only up to a point. The first partial derivatives of Eq. (15.6) from which the best-response functions in Eqs. (15.14) and (15.15) were derived indicate that each division's share of the profits is a decreasing function of its level of output. Division 1's share of the profits, for example, begins to decline once its output level exceeds $(2 + cx_2)$, unless division 2 expands output as well. Finally, profit increases with the degree of strategic complementarity from zero to, but not including, 0.5 (why?).

The conflict between individual and group incentives was encountered earlier in our discussion of coalition games. The main difference here is that the Nash equilibrium strategy profile was derived by iteratively eliminating dominated strategies. The reasons for the conflict are due to differences between personal and joint costs, and because the divisions receive only a share of the total benefits. Since both divisions expend less effort, the firm is worse off.

Solved Exercise

Suppose in the above example that the degree of complementary effort is $c = 0.2$.

a. *What is the equilibrium strategy profile and total profit from double marginalization?*
b. *What is the equilibrium strategy profile when divisions maximize joint profits?*

Solution

a. *From condition (15.12), the equilibrium strategy profile from double marginalization is*

$$\left\{\left(\frac{2}{1-c}\right), \left(\frac{2}{1-c}\right)\right\} = \{2.5, 2.5\}. \tag{15.19}$$

From Eq. (15.6), the firm's total profit from this strategy profile is

$$\pi^* = \frac{4}{(1-0.2)^2} = 6.25. \tag{15.20}$$

b. From condition (15.16), the equilibrium strategy profile when divisions maximize joint profits is

$$\left\{\left(\frac{2}{1-2c}\right),\left(\frac{2}{1-2c}\right)\right\} = \{3.33, 3.33\}. \tag{15.21}$$

From Eq. (15.17), the firm's total profit at this strategy profile is

$$\hat{\pi} = \frac{4}{[1-2(0.25)]^2} = 8. \tag{15.22}$$

Practice Exercises

15.1 How would you describe the relationship between the firm's products when $c < 0$?

15.2 A firm is organized into two divisions. The goods produced by these divisions are demand complements. The profit equations of the two divisions are

$$\pi_1(x_1, x_2) = 4x_1 + cx_1x_2 - x_1^2; \tag{PE15.2a}$$

$$\pi_2(x_1, x_2) = 4x_2 + cx_1x_2 - x_2^2. \tag{PE15.2b}$$

where x_1 and x_2 are the outputs of goods 1 and 2, respectively, and $c = 0.5$ is the degree of strategic complementarity.

a. What is the equilibrium strategy profile and total profit if each division is treated as a quasi-independent profit center?

b. What is the equilibrium strategy profile and total profit when the divisions maximize joint profits?

c. For what range of values of c are the profits from joint profit maximization greater than firm profits from double marginalization?

15.3 A firm's profit depends on the separate and joint efforts of two partners, which is given by the equation

$$\pi(x_1, x_2) = 4x_1 + 4x_2 + 4cx_1x_2. \tag{PE15.3}$$

Suppose that the partners incur private opportunity costs of x_1^2 and x_2^2, and share equally in the firm's profits.

a. Under what conditions are the partners strategic complements?

b. What is the equilibrium strategy profile if the partners maximize their share equally in the firm's profits? What is the firm's profit?

c. What equilibrium strategy profile will maximize the partners' joint profits? What is the firm's profit?

d. For what values of c will the firm's profit from joint maximization exceed the profit from double marginalization?

PART II

DYNAMIC GAMES WITH COMPLETE AND PERFECT INFORMATION

CHAPTER SIXTEEN

Game Trees

Introduction

In this and the next two chapters we explore the basic elements of dynamic games under conditions of complete and perfect information. Also referred to as *sequential-move games*, dynamic games differ from static games in that the players alternate moves. Except for the first mover, players observe a rival's countermoves and then use that information to update their strategies whenever possible.

Game Trees

The diagrammatic representation of a sequential-move game is referred to as the *extensive form* or, more popularly, a **game tree**. A game tree summarizes the players, the information available at each stage, the order of the moves, and the payoffs from alternative strategy profiles. In a dynamic game, a strategy is the complete description of a player's moves at each stage.[1]

In addition to complete information, we also assume **perfect information,** which means players know their precise location in the game tree. This is because they know with certainty how other players have already moved. This is in contrast to the static games previously examined in which players do not know the moves of the other players until all moves have been made. In dynamic games with imperfect information, players are not sure about their location in the game tree because they are uncertain about how another player has already moved. Dynamic games with imperfect information will be examined later in the text.

Game trees illustrate the strategic interaction of two or more players who alternate moves at branching points called **decision nodes**. In the game tree depicted in Fig. 16.1, for example, decision nodes are depicted as small, closed circles. Each

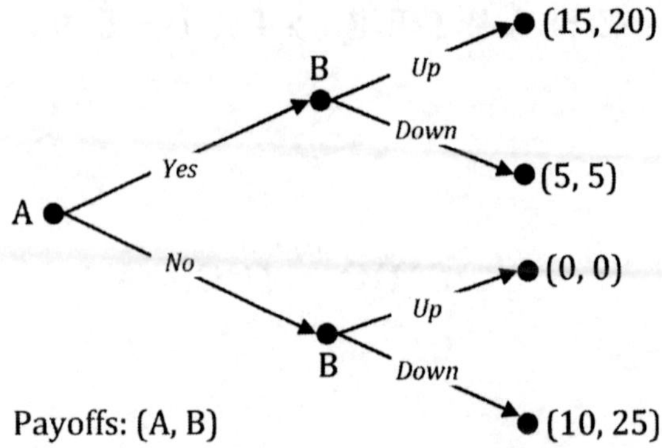

Fig. 16.1.

decision node indicates the player who moves at that point in the game. In this game, the first decision is made by player A. This decision node is called the **root of the game tree**.

The strategy choices of the designated player at each decision node are represented by arrows called **branches**. All decision nodes after the initial decision node are called *subroots*. The small, closed circles at the far right are the "leaves" of the game trees, which are referred to as **terminal nodes**. This is where the game ends. To the right of the terminal nodes are the players' payoffs. The first entry in the parentheses is the payoff to player A and the second entry is the payoff to player B. A game tree is the complete collection of decision nodes, branches, terminal nodes, and payoffs.

The order of play in Fig. 16.1 is from left to right. In the first stage, player A moves *yes* or *no*. In the second stage, B responds by moving *up* or *down*. The ideal order of play for A is *yes* followed by *up*, which results in payoffs of (15, 20). The ideal order of play for B is *no* followed by *down*, which results in payoffs of (10, 25). Since A moves first, the sequence *yes* followed by *up* is an equilibrium strategy profile.

Subgame Perfection

The idea of equilibrium in sequential-move games owes much to the work of Reinhard Selten (1975), who was awarded the 1994 Nobel Prize in economics for his pioneering work in game theory along with John Nash and John Harsanyi. Selten formalized the concept of a **subgame**, which is any subset of branches and decision nodes in a game tree. Selten's important contribution was to recognize that each subgame is itself a game.

The first decision node of a subgame is called a **subroot**. Every decision node is the subroot of a new subgame. A subgame has three basic characteristics. First, all subgames have the same players as the entire game itself, even if some of these players do not move. Second, a subgame includes the subroots, branches, decision nodes, and terminal nodes of all subsequent subgames. Since every game is a subgame of itself, the game beginning at the root of the game tree is referred to as a **trivial subgame**. Any subgame that does not include the root of the game tree is called a **proper sub-**

game. Finally, since a trivial subgame includes all proper subgames, the payoffs of all subgames are included among the payoffs for the entire game.

Once a subgame begins, a player can neither turn back nor exit. Each subgame is self-contained and has its own **subgame equilibrium**. Thus, there are at least as many subgame equilibria as there are subgames. A unique equilibrium for the entire game, which is found among the complete collection of subgame equilibria, is called a **subgame-perfect equilibrium**.

The concept of a subgame is illustrated in Fig. 16.2, which replicates Fig. 16.1. This game has three subgames. The first subgame, which begins at the root of the game tree (S_1), is a trivial subgame. Two proper subgames begin at decision nodes (subroots) S_2 and S_3. The subgame that begins at subroot S_2, which is enclosed by the dashed, rounded rectangle, has two terminal nodes, T_1 and T_2, with payoffs of (15, 20) and (5, 5), respectively.

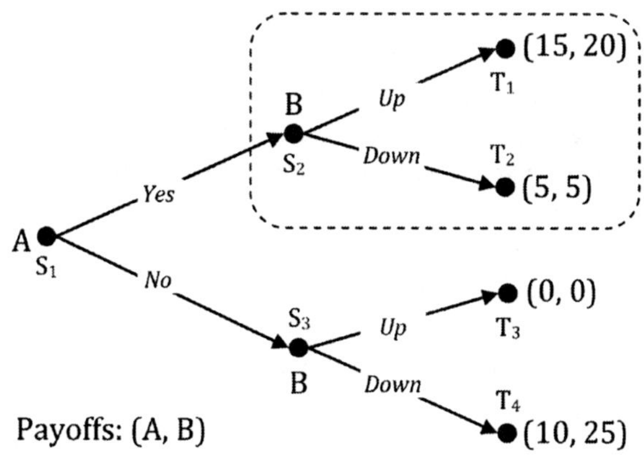

Fig. 16.2.

Backward Induction

The most commonly used method for identifying a subgame-perfect equilibrium, which was developed by Reinhard Selten (1965), is **backward induction**. The rationale underlying this solution algorithm is to determine a rival's best response to a move and use that information to formulate a strategy that will result in the best payoff. In other words, a player should project forward and reason backward.

Backward induction involves five steps:

1. Start at the terminal nodes. Trace each branch to its decision node. A *basic decision node* has multiple branches that all lead to terminal nodes. A *trivial decision node* has only one branch that leads to a terminal node. A *complex decision node* has at least one branch that leads to another decision node.
2. Compare the payoffs for the player at the decision nodes identified in step 1. The best response leads to the preferred payoff. All other branches should be pruned away.

3. Move backwards to the next decision node. Identify the branches that lead to that player's best payoff. Prune away all other branches from that decision node.
4. Repeat steps 1–3 until the root of the tree has been reached.
5. From the root of the game tree identify which of the remaining branches leads to the best payoff for the first mover. That sequence branches is the subgame-perfect equilibrium.

The extensive form, which may also be used to analyze static games, has several advantages over the normal form. For one thing, game trees and backward induction allow us to consider games involving more than three players with multiple strategies and stages. What is remarkable about backward induction in games with complete and perfect information is that it will almost always result in a unique Nash equilibrium. A possible exception to this is when a player is indifferent between two or more moves at a basic decision node.

Solved Exercise

Nexus Corporation is planning to introduce a new online computer game. It is considering two marketing strategies. An expensive, Madison Avenue–style slick *strategy will result in high sales the first year, but substantially lower sales in the second year as the market becomes saturated. The inexpensive word-of-mouth* simple *strategy will result in low sales the first year, but much higher sales in the second year. Regardless of marketing strategy, no significant sales are anticipated after the second year.*

A potential problem for Nexus is that Mobius Company developing a legal clone that is a near-perfect substitute, but with a sufficiently different computer code so that it cannot be sued for copyright infringement. It will take a year for Mobius to develop its clone. If Mobius enters the market, the firms will split sales in the second year. The present values of the payoffs from alternative strategy profiles are summarized in Fig. 16.3. If larger payoffs are preferred, what is the subgame-perfect equilibrium?

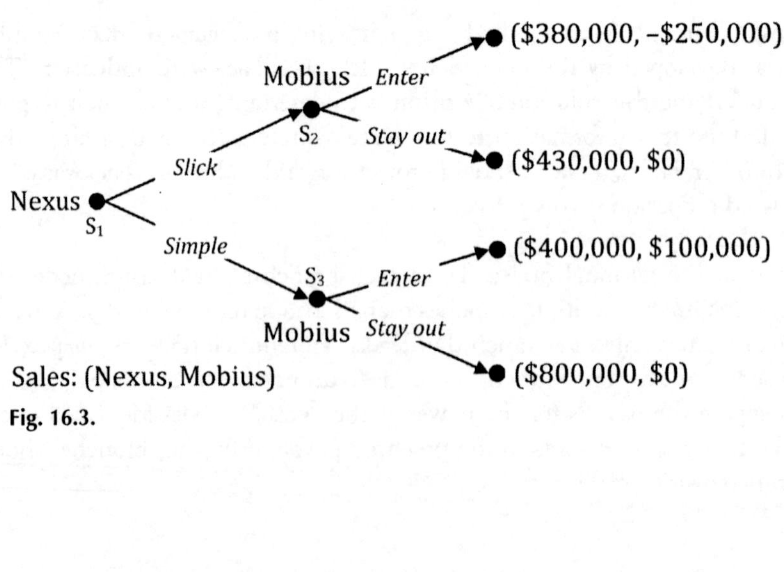

Fig. 16.3.

Solution

Starting at each terminal node, move to the Mobius decision nodes. Each of these decision nodes is basic since all of the branches lead to terminal nodes. If Nexus moves slick, *the best response by Mobius is* stay out *since $0 is greater than –$250,000. The* enter *branch should be removed. If Nexus moves* simple, *the best response by Mobius is* enter *since $100,000 is greater than $0. The* stay out *should also be pruned away. The resulting extensive form game is depicted in Fig. 16.4. From the root of the game tree, Nexus should move* slick *since $430,000 is greater than $400,000. The subgame-perfect equilibrium strategy profile is for Nexus to move slick and for Mobius to* stay out. *When there is no risk of confusion, we will identify the subgame-perfect equilibrium as the sequential collection of branches, that is, {Slick → Stay out}.*

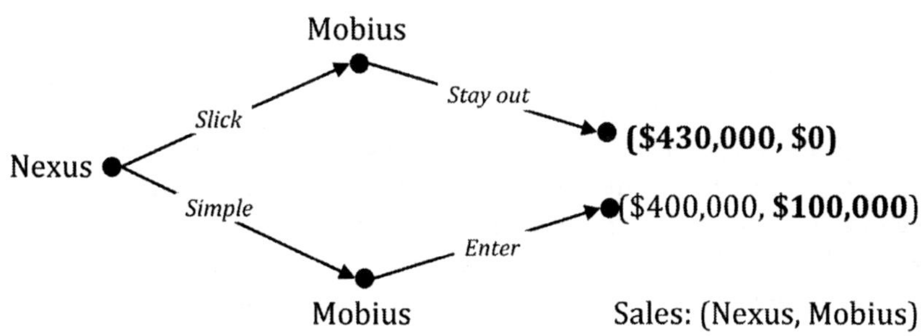

Fig. 16.4.

Credible Threats

The purpose of a strategic move is to alter a rival's behavior. A threat is an example of a strategic move, but only if it is credible. A threat is credible if it is in a player's best interest to follow through with the threat. Otherwise, it is an empty threat. We can use backward induction to identify credible and empty threats in sequential-move games. The reader will recall that the subgame-perfect equilibrium in Fig. 16.1 is {*Yes* → *Up*} with payoffs of (15, 20). The best outcome for B, however, is the strategy profile {*No* → *Down*} with payoffs of (10, 25). How can B convince A to move *no* instead of *yes*?

Suppose that B threatens to move *down* regardless of the move made by A. If A believes that B's threat is credible, it will be in A's best interest to move *no*. B's threat, however, is not credible. If B is rational and A moves *yes*, B's best response is *up*.

The importance of credibility can also be demonstrated with the help of Fig. 16.3. The subgame equilibrium beginning at S_2 is {*Stay out*}. The subgame equilibrium beginning at S_3 is {*Enter*}. The subgame-perfect equilibrium must include one of these subgame equilibria. Suppose that Mobius tries to persuade Nexus to move *simple* by announcing its unconditional intention to *enter*. Without some reinforcing measure to bolster the threat, Mobius will interpret Nexus' announcement as an empty threat. Nexus will adopt a slick advertising strategy and a rational Mobius will stay out.

In general, empty threats should be eliminated whenever we require that equilibrium strategies apply to any subgame. The reasoning is straightforward. The only move that matters is the move with the greatest payoff. This is precisely the move that is selected using backward induction, which automatically eliminates all empty threats. For this reason, equilibria for sequential-move games with complete and perfect information are said to be *subgame perfect*.

First-Mover Advantage

Anyone who is familiar with the game of chess knows that white has the early advantage since it moves first. This is because black is immediately placed on the defensive. This **first-mover advantage** is a feature of sequential-move games. A first-mover is a player who can irrevocably commit to a strategy before any other player. Other things being equal, a first-mover will enjoy a payoff that is no worse than the payoff in a static game—and could be better.

The first-mover advantage can be illustrated with the well-known chicken game, which is depicted in Fig. 16.5. In this game two teenagers, Dean and Brando, challenge each other to a test of nerves by driving their cars directly at each other. The driver who swerves loses his car to the driver who does not swerve. If neither swerves they crash, which could result in serious injury or death. At the very least, their cars will be wrecked. If both swerve they will walk away unharmed, with the possible exception of bruised damaged egos.

		Brando	
		Swerve	*Commit*
Dean	*Swerve*	(−1, −1)	**(0, 10)**
	Commit	**(10, 0)**	(−10, −10)

Payoffs: (Dean, Brando)

Fig. 16.5.

This game has two Nash equilibria in pure strategies: {*Commit*, *Swerve*} and {*Swerve*, *Commit*}. The strategy profile {*Commit*, *Swerve*} is preferred by Dean and {*Swerve*, *Commit*} is preferred by Brando. The key to turning this game to a driver's advantage is to make an irrevocable first move. To see how, suppose as they race toward each other that Dean rips out his steering wheel and tosses it out the window for Brando to see. By making a credible commitment not to swerve, Dean has transformed a static game into the sequential-move game depicted in Fig. 16.6. Using backward induction, the subgame-perfect equilibrium is {*Commit* → *Swerve*} and Dean is rewarded with his preferred payoff.

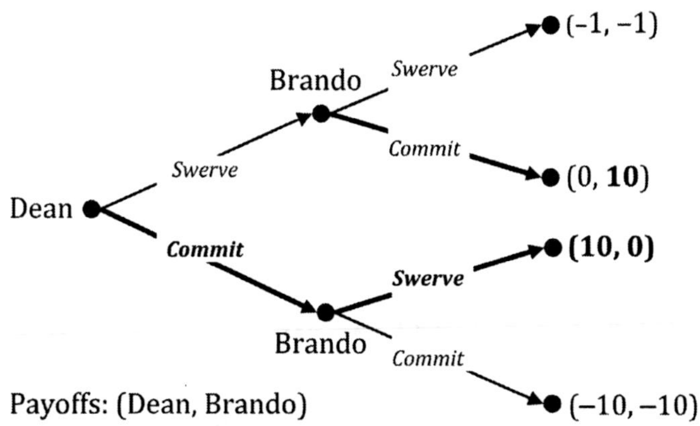

Fig. 16.6.

It is sometimes argued that a key to success in business is to be the first to market a new product or innovation. Amazon.com, Bayer, Coca-Cola, IBM, Intel, Kodak, Microsoft, Standard Oil, and Xerox are well-known examples. Being first, however, does not guarantee success. It is commonly believed that Proctor & Gamble (P&G) was first to introduce the re-sealable, disposable diaper (Pampers) in 1961. In fact, this distinction belongs to Kimberly Clark, which introduced the now defunct Chux diapers. P&G succeeded because it offered a lower cost product with wider market appeal. A similar story applies to video recorders, which were first developed in 1956 by California-based Ampex. Twenty years later, this market was dominated by Japan's Matsushita and Sony.[2]

There are several possible explanations for why first movers failed to maintain their leadership position. Having a first-mover advantage only says that a player who can commit to a strategy first, and does not falter, will enjoy a payoff that is no worse than if all players moved at the same time. Unfortunately, technically innovative managers are not always good game players. Failure to correctly project forward and reason backward frequently leads to lower profits, disappointed stakeholders, and corporate failure.

Solved Exercise

Consider the one-time, static game in Fig. 16.7.

a. Does this game have a Nash equilibrium?
b. What is the subgame-perfect equilibrium if player A moves first? Does A enjoy a first-mover advantage?

		B		
		B1	*B2*	*B3*
	A1	(4, 2)	(2, 1)	(2, 1)
A	*A2*	(3, 1)	(4, 1)	(3, 3)
	A3	(5, 4)	(2, 5)	(2, 1)

Payoffs: (A, B)

Fig. 16.7.

Solution

a. *The Nash equilibrium strategy profile for this game is {A2, B3} with payoffs of (3, 3).*
b. *The extensive form of this game when A moves first is depicted in Fig. 16.8. Using the method of backward induction, the subgame-perfect equilibrium for this game is {A1 → B1} with payoffs of (4, 2). By moving first, A has improved his payoff from 3 to 4. Player A has a first-mover advantage.*

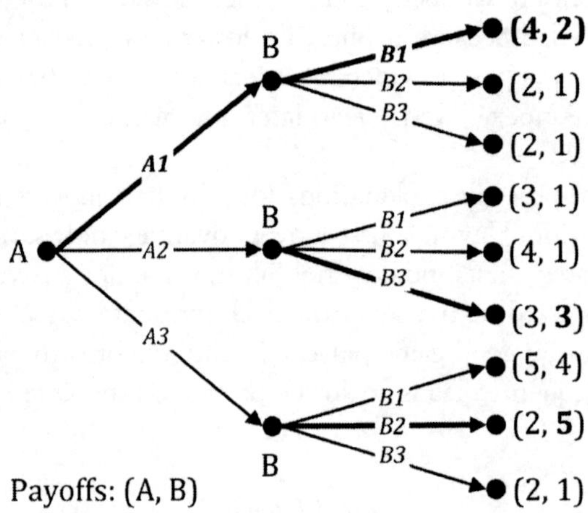

Fig. 16.8.

Entry Deterrence

Predatory (limit) pricing occurs when an incumbent firm charges a price that is low enough to deter entry into the market by potential competitors. For predatory pricing to succeed in the long run, potential rivals must believe that the incumbent's threat to slash prices is credible, which may depend on several factors, such as whether the incumbent's pockets are deep enough to wage a protracted price war.

The use of threats and the first-mover advantage to deter entry by potential rivals is depicted in Fig. 16.9. In this game, David is a start-up software company and Goliath is the established industry leader. David has developed a faster, more efficient operating system, which it believes can significantly eat into Goliath's market share. If David *enters* and Goliath adopts a *competitive* pricing policy, the present values of the firms' profits are $500,000 and $2 million, respectively. Can Goliath deter David's entry by threatening *predatory* pricing?

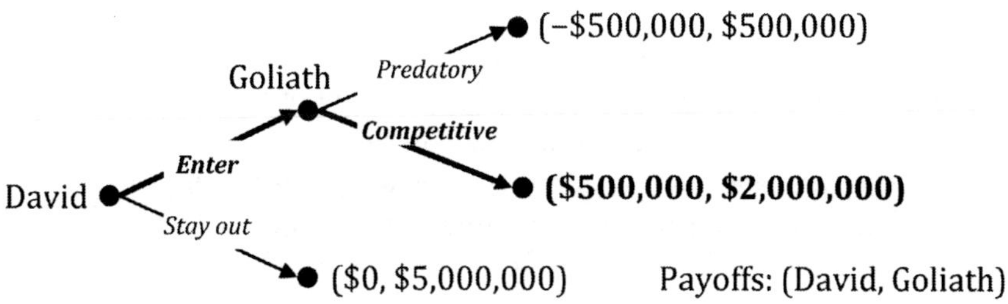

Fig. 16.9.

Goliath's threat to slash prices is a bluff. If David *enters*, Goliath will earn $500,000 by adopting a *predatory* pricing strategy, compared with $2 million by being *competitive*. Using the method of backward induction, the subgame-perfect equilibrium strategy profile for this game is {*Enter* → *Competitive*}.

Analytically, the problem for Goliath is that David is the first-mover. If Goliath can reverse these roles, it may be possible to turn the game to its advantage. Suppose that the CEO of Goliath leaks an internal memo indicating that it will file a lawsuit claiming copyright infringement if David enters the market. The lawsuit will cost each company $1 million in legal fees and court costs, which Goliath can easily afford.

Start-up companies typically need to get their product to market quickly to generate much-needed cash flows. Although David's lawyers are dubious of the merits of the lawsuit, the company is financially unprepared to weather such a legal challenge. The effect of Goliath's preemptive lawsuit strategy is depicted in Fig. 16.10. Using

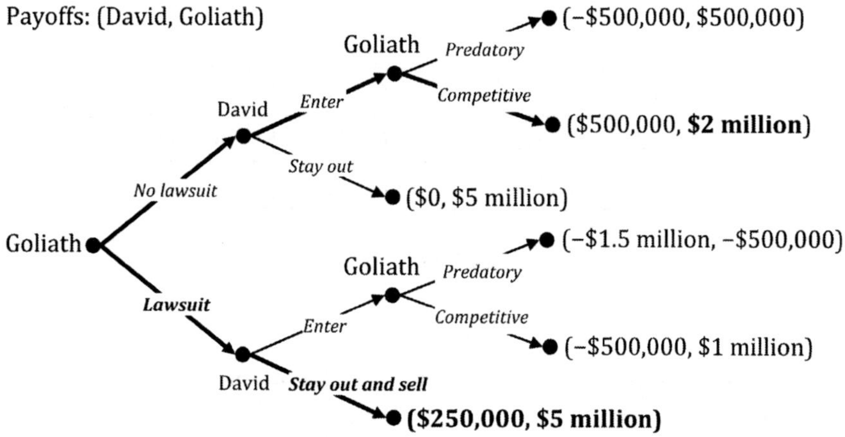

Fig. 16.10.

backward induction, the subgame-perfect equilibrium is for Goliath to file a lawsuit, and for David to stay out of the market and sell the copyrights to its new operating system to Goliath for $250,000.

Practice Exercises

16.1 Why is a sequential-move game with complete and perfect information using backward induction subgame perfect?

16.2 Explain how backward induction eliminates empty threats.

16.3 Consider the following one-time, static game.

a. Does this game have a unique Nash equilibrium strategy profile?

b. What is the subgame-perfect equilibrium if A moves first?

			B	
		B1	*B2*	*B3*
	A1	(4, 4)	(4, 3)	(2, 3)
A	*A2*	(1, 2)	(3, 4)	(2, 5)
	A3	(5, 1)	(3, 2)	(3, 1)

Payoffs: (A, B)

Fig. PE16.3.

16.4 Consider the following one-time, static game.

a. Does this game have a unique Nash equilibrium strategy profile?

b. What strategy profile results in the best outcome for both players?

c. What is the subgame-perfect equilibrium if A moves first?

d. Does A have a first-mover advantage?

			B	
		B1	*B2*	*B3*
	A1	(4, 4)	(4, 3)	(2, 3)
A	*A2*	(1, 2)	(3, 4)	(2, 5)
	A3	(5, 1)	(3, 2)	(3, 3)

Payoffs: (A, B)

Fig. PE16.4.

16.5 Consider the centipede game depicted in the following figure in which two players have the opportunity to take a larger portion of an ever-increasing payoff. What is the subgame-perfect equilibrium?

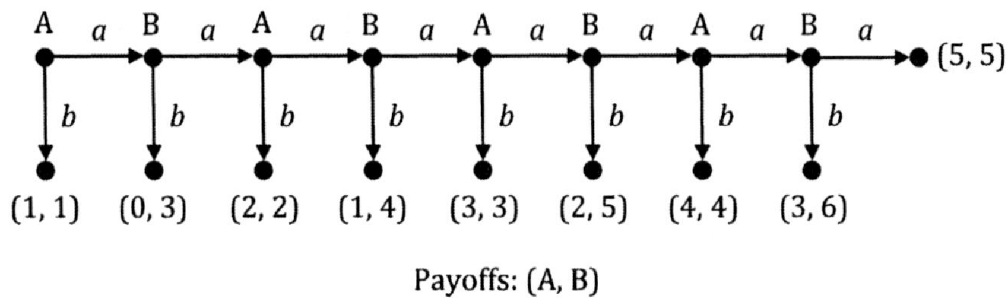

Fig. PE16.5.

16.6 Consider the following centipede game. What is the subgame-perfect equilibrium?

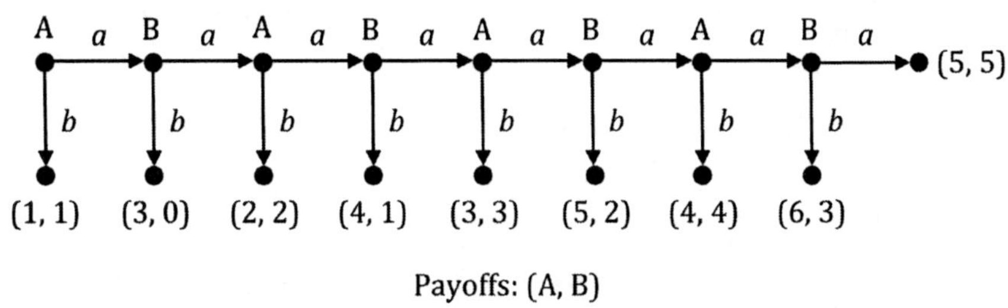

Fig. PE16.6.

16.7 Tom, Dick, and Harry are playing the sequential-move game in the following figure. What is the subgame-perfect equilibrium?

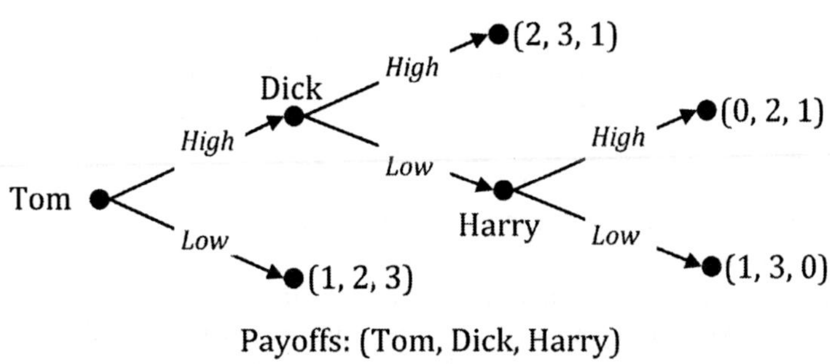

Fig. PE16.7.

16.8 Tom, Dick, and Harry are playing the sequential-move game in the following figure. What is the subgame-perfect equilibrium?

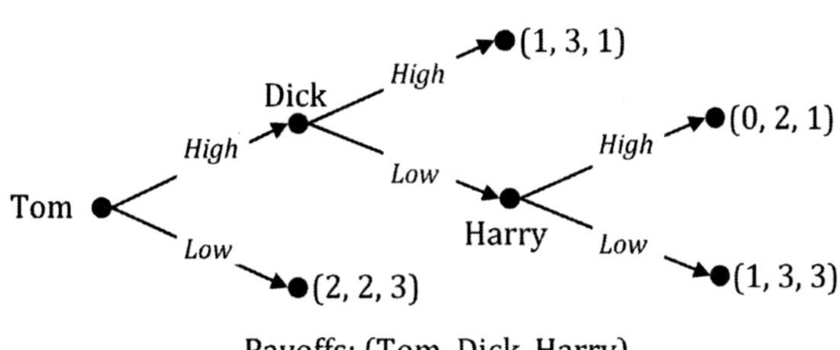

Fig. PE16.8.

16.9 What is the subgame-perfect equilibrium in the following sequential-move game?

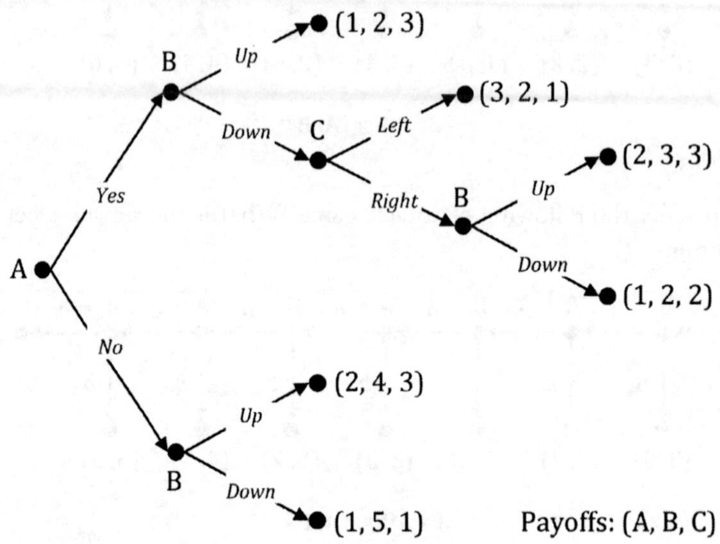

Fig. PE16.9.

Notes

1. Game trees are similar to *decision trees*, which are used by managers to compare expected values and risks of alternative undertakings to achieve some objective.

2. This pattern of first-mover being outdone by imitators has been repeated time and again. A study by Tellis and Golder (1996) found that in more than fifty markets, first-movers were industry leaders only 10 percent of the time. Moreover, current industry leaders on average entered the market thirteen years after the first-mover. One explanation for this is that industry leaders have little incentive to risk their market share by bringing an unproven product to market. In general, their dominant strategy is to "follow the pack." IBM, for example, is not known for product innovation, but rather for its ability to standardize technology for mass-market consumption. In fact, most new ideas came from such start-up companies as Sun Microsystems and Apple because risky innovations were their best hope of capturing market share.

CHAPTER SEVENTEEN

A Dynamic Oligopoly Game

Introduction

In this chapter, we will discuss a dynamic variation of the static Cournot output-setting game discussed in chapter 12. Proposed by Heinrich von Stackelberg (1934), the **Stackelberg model** is similar to the Cournot model, but with a critical difference. In the Cournot model the firms determined output simultaneously. In the Stackelberg model, output is determined sequentially. As before, we assume complete and perfect information.

Stackelberg Model

Once again, consider an industry consisting of two profit-maximizing firms. Firm 1 decides how much to produce, which is observed by firm 2. Firm 2 then decides how much it should produce. The firm that commits to an output strategy first is referred to as a **Stackelberg leader**. This is followed by the output decision of the second firm, which is referred to as the **Stackelberg follower**.

The important thing about the Stackelberg model is that only the Stackelberg follower has a best-response function, which is known by the Stackelberg leader. By projecting forward and reasoning backward, the Stackelberg leader exploits the Stackelberg follower's best response to maximize its own profit. In the Stackelberg model, the Stackelberg leader has a first-mover advantage.

To illustrate the Stackelberg model, we will assume the same demand and cost used in our discussion of the Cournot model, which are summarized by Eqs. (12.1) and (12.4). The main difference between the Cournot and Stackelberg models is the timing of the firm's output decisions. In the Stackelberg model, firm 1 chooses an output level Q_1^* that maximizes its profit, which firm 2 incorporates into its best-response function to determine its profit-maximizing level of output $Q_2*(Q_1*)$. Firm 2's best-response function is given by Eq. (12.8). Whereas the Nash equilibrium strategy profile in the Cournot output-setting game is $\{Q_1*(Q_2*), Q_2*(Q_1*)\}$, the subgame-perfect equilibrium for the sequential-move Stackelberg game is $\{Q_1*, Q_2*(Q_1*)\}$.

To determine firm 1's profit-maximizing output, substitute firm 2's best-response function into Eq. (12.2), which gives us

$$TR_1 = 17.5Q_1 - 0.25Q_1^2. \tag{17.1}$$

From Eq. (17.1), firm 1's marginal revenue is

$$MR_1 = 17.5 - 0.5Q_1. \tag{17.2}$$

Equating firm 1's marginal revenue to MC_1 = \$10, the profit-maximizing level of output for firm 1 is $Q_1* = 15$. From Eq. (12.8), the profit-maximizing output level for firm 2 is $Q_2* = 7.5$. Substituting total industry output of 15 + 7.5 = 22.5 into Eq. (12.1), the market-clearing price is $P*$ = \$13.75. Thus, the Stackelberg leader earns a profit of π_1* = (\$13.75 – \$10)15 = \$56.25, while the Stackelberg follower earns a profit of π_1* = (\$13.75 – \$10)7.5 =\$28.125. The Stackelberg leader enjoys a first-mover advantage because it earns a greater profit than when both firms move simultaneously as in the Cournot model (\$50).

Other things being equal, a first-mover realizes a payoff that is no worse than if all players were to move simultaneously. By committing to an output level first, the Stackelberg leader in the above example not only enjoyed greater profits than the Stackelberg follower, but earned profits that were greater profits than when both firms set their output levels simultaneously, as was the case in the Cournot output-setting model. The Stackelberg leader's first-mover advantage might be traced to several sources, including superior management, a cutting-edge research and development program that enables the firm to bring new or improved products ahead of its competition, a nexus of long-term contracts with input suppliers that enables the firm to bring its product to market more quickly.

The existence of long-term supplier contracts, however, could work to firm 1's disadvantage by making it difficult to respond quickly to changes in market conditions. Frequent shifts in consumer tastes or rapidly emerging technology could make the leader vulnerable to challenges from rivals seeking to supplant its leadership position. This suggests that the kind of Stackelberg competition described above is more likely to occur in mature industries in which product demand is stable, or when changes in product technology are infrequent and predictable. Stackelberg competition is less likely in situations in which predicting changes in consumer tastes is akin to hitting a moving target, or where product development is rapid and unpredictable, such as in the fashion or consumer electronics industries.

Solved Exercise

Consider an industry consisting of two firms producing an identical product. The market demand and total cost of these firms are

$$P = 10 - 2(Q_1 + Q_2); \quad (17.3)$$

$$TC_1 = 4Q_1; \quad (17.4)$$

$$TC_2 = 2Q_2. \quad (17.5)$$

Assume that firm 1 is the Stackelberg leader and firm 2 the Stackelberg follower.

a. *What are the equilibrium price, output levels, and profits for each firm?*
b. *What effect does the Stackelberg follower's lower marginal cost have on the Stackelberg leader's first-mover advantage?*

Solution

a. *The best-response function of the Stackelberg follower is*

$$Q_2 = 2 - 0.5Q_1. \quad (17.6)$$

Substituting Eq. (17.6) into firm 1's total revenue equation we obtain

$$TR_1 = 6Q_1 - Q_1^2. \quad (17.7)$$

From Eq. (17.6), firm 1's marginal revenue equation is

$$MR_1 = 6 - 2Q_1. \quad (17.8)$$

Equating marginal revenue with MC_1 *= $4, firm 1's profit-maximizing level of output is* $Q_1^* = 1$. *Substituting this result into Eq. (17.6), firm 2's profit-maximizing level of output is* $Q_2^* = 1.5$. *Total industry output is* $Q_1^* + Q_2^* = 2.5$. *From Eq. (17.3), the market clearing price is* $P^* = \$5$. *Firm 1 will earn a profit of* $\pi_{1*} = (\$5 - \$4)1 = \$1$. *Firm 2 will earn a profit of* $\pi_{2*} = (\$5 - \$2)1.5 = \$4.5$.

b. *In spite moving first, firm 1 produces less and earns a lower profit than firm 2. Firm 2's lower marginal cost has negated firm 1's first-mover advantage.*

Practice Exercises

17.1 How does the Stackelberg duopoly model modify the Cournot duopoly model?

17.2 What is the first-mover advantage in the Stackelberg game? Apart from the examples cited in the chapter, can you think of other sources of a first-mover advantage?

17.3 Suppose that the demand equation for the output of the industry is

$$P = 45 - 5(Q_1 + Q_2). \qquad \text{(PE17.3a)}$$

The total cost equations are

$$TC_1 = 5Q_1; \qquad \text{(PE17.3b)}$$

$$TC_2 = 5Q_2. \qquad \text{(PE17.3c)}$$

Suppose that firm 1 is a Stackelberg leader and firm 2 is a Stackelberg follower.

a. What are the equilibrium price, output levels, and profits for each firm?
b. Does the Stackelberg leader have a first-mover advantage?

17.4 Suppose that the demand equation for the output of the industry is

$$P = 45 - 5(Q_1 + Q_2). \qquad \text{(PE17.4a)}$$

The total cost equations are

$$TC_1 = 3Q_1; \qquad \text{(PE17.4b)}$$

$$TC_2 = 5Q_2. \qquad \text{(PE17.4c)}$$

Suppose that firm 1 is a Stackelberg leader and firm 2 is a Stackelberg follower.

a. What are the equilibrium price, output levels, and profits for each firm?
b. Does the Stackelberg leader have a first-mover advantage?

CHAPTER EIGHTEEN

Bargaining

Introduction

Whether in business, politics, law, international relations, or everyday life, bargaining is an important aspect of everyday life. Whether negotiating with your boss for an increase in salary and benefits, haggling over the price of a new car, hammering out an agreement with suppliers over the prices of raw materials or the specifications of components, it is useful to think of bargaining as a process whereby rivals negotiate over the division of something of value. In this chapter, we will apply the principles of game theory to deepen our understanding of the bargaining process.

The Bargaining Problem

Trade is the process whereby individuals exchange goods or services. If these transactions are voluntary, both parties benefit. All market transactions involve implicit or explicit contracts. **Bargaining** is a process of negotiating the terms of a contract. If either party is not satisfied, negotiations fail and the transaction does not take place. If negotiations are successful, how will the gains from trade be distributed?

In general, there are two questions that must be answered when analyzing bargaining scenarios. First, what are the bargaining rules? Second, what are the payoffs when the players reach, or fail to reach, an agreement? In most retail establishments, for example, the seller posts a fixed asking price. The customer must decide whether to accept or reject this price. This is an example of a take-it-or-leave-it bargaining rule. In the case of collective bargaining, union representatives propose a wage and benefit package. Management can accept or reject. If management rejects, it can wait for the union to modify its proposal, or it can make a counteroffer. In some cases, the order of play is determined by custom or law. In other cases, strategic considerations may determine the sequencing of the bargaining process.

Bargaining games with multiple Nash equilibria are not uncommon. For this reason, we want to understand when it is in the parties best interest to cooperate. In general, the best outcome is the proposed distribution of a disinterested third party. A player should never accept a Pareto inefficient distribution. That is, players should cooperate when no player can be made better off without making some other player worse off.

To make the discussion of the bargaining process more concrete, suppose that two individuals are negotiating over the sale of a used car. If an agreement is reached, the sale involves a monetary transfer (t) from the buyer to the seller. If we assume that a player's utility is additive and transferrable, increasing the offer price by a dollar simultaneously reduces the buyer's utility and increases the seller's utility by the same amount. Thus, the utility functions of the buyer (u_B) and the seller (u_S) may be written

$$u_B = v_B - t; \tag{18.1}$$

$$u_S = t - v_S. \tag{18.2}$$

The term v_B is the **buyer's reservation price**, which is the maximum price that the buyer is willing to pay. The term v_S is the **seller's reservation price**, which is the minimum price that the seller is willing to accept. An efficient bargaining outcome maximizes the sum of the players' utility ($u_B + u_S$). Since the amount t represents an equal monetary transfer, it has no effect on the players' joint payoff.

The objective of the players is to maximize their respective shares of the **bargaining surplus**, which depends on a variety of factors, including the players' reservation prices, negotiating skills, eagerness to reach an agreement, and general contracting environment. The bargaining surplus is defined as the difference between the buyer's and the seller's reservation prices, that is

$$v_B - v_S = u_B + u_S. \tag{18.3}$$

For an agreement to be possible, it must be the case that $v_B \geq v_S$. The bargaining process involves the distribution of a positive bargaining surplus. Suppose, for example, that the buyer's reservation price is v_B = \$500 and the seller's reservation price is v_S = \$400. The bargaining surplus is $v_B - v_S$ = \$100. If the buyer and seller agree to a transfer price of t = \$425, the utility received by the buyer is $v_B - t$ = \$75 and the utility received by the seller is $t - v_S$ = \$25. The sum of the player's utility from a bargaining agreement is $u_B + u_S$ = \$100.

Ultimatum Bargaining

The concepts presented in the previous section can be used to illustrate one of the simplest of all bargaining scenarios—the **ultimatum game**. The ultimatum game is a take-it-or-leave-it bargaining scenario. In the first stage of this two-stage, sequential-move game, the buyer makes an offer, which includes a share of the bargaining surplus. In the second stage, the seller either accepts or rejects the offer, at which time the game ends.

In the used-car bargaining game, for example, suppose the buyer makes a take-it-or-leave-it offer of $400. If the offer is accepted, the buyer receives the entire $100 bargaining surplus. If an offer of $500 is accepted, the seller receives the full amount of the bargaining surplus. If an offer of $450 is accepted, the buyer and seller split the bargaining surplus. The buyer will never offer more than $500. The seller will never accept an offer less than $400. How much should the buyer offer and what will be the distribution of the bargaining surplus?

To generalize the discussion, suppose that the bargaining surplus is viewed as a monetary pie. This two-stage game is depicted in Fig. 18.1. How large a slice should the buyer offer the seller? In the first stage, the buyer offers the seller a share $0 \leq \beta \leq 1$. If the buyer accepts, the payoffs are $(1 - \beta, \beta)$. If the buyer rejects, the payoffs are (0, 0). This game has an infinite number of subgames, each corresponding to an infinite number of possible distributions.

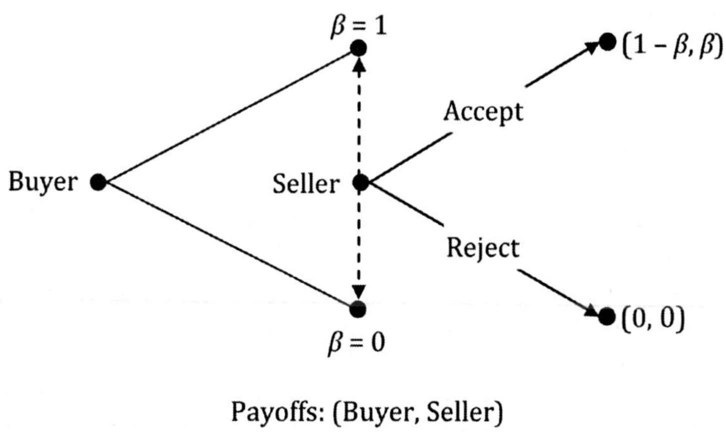

Fig. 18.1.

Because the payoffs are in money and utility is transferrable, the bargaining set for this game is depicted as the downward sloping straight line *bb* in Fig. 18.2. The subgame-perfect equilibrium can be found by considering any offer of $0 \leq \beta \leq 1$. The seller has two sequentially rational strategies: (1) accept all offers, or (2) accept all offers of $0 < \beta \leq 1$ and reject the offer when $\beta = 0$. These are the only strategies that define the equilibrium for a proper subgame. This is because the default outcome (disagreement point) *d* lies below every point along the bargaining set *bb*. If the seller rejects the offer, the payoffs are (0, 0). Thus, $\beta = 0$ cannot lead to a subgame-perfect equilibrium. On the other hand, any positive offer of $0 < \beta \leq 1$ will result in a positive payoff for both players since something is better than nothing. This implies that the buyer should offer the smallest possible share, which should be accepted.

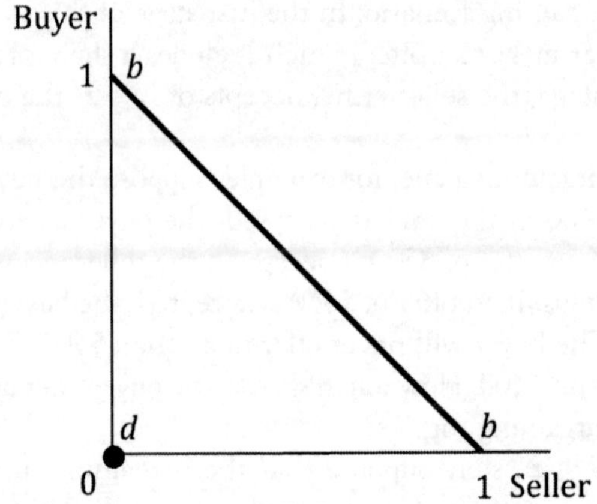

Fig. 18.2.

On its surface, the ultimatum game underscores the power of the player making the take-it-or-leave-it offer in the first round. As the following discussion of the ultimatum paradox illustrates, however, decisions based on rational, self-interested behavior are not always what is observed in reality.

Ultimatum Paradox

Analysis of what came to be called the ultimatum paradox began with a two trial experiment by Werner Güth, Rolf Schmittberger, and Bernd Schwarze (1982) in which members of a group of "proposers" (givers) were paired with members of another group of "responders" (takers). They found that the givers' offers were inconsistent with the predictions of rational behavior that something, however small, is better than nothing. In the first part of their experiment, givers offered takers an even split 7 of 21 times, with a mean offer of 37 percent. In only one instance was the proposed distribution rejected. When Guth et al. repeated their experiment a week later with the same test subjects, the proposed offers were somewhat less generous, but still significantly greater than the small shares predicted by rational behavior. Takers were offered a 50-50 split in two instances; the mean offer was 32 percent, and 5 of 21 offers were rejected. In a separate experiment, a different group of 37 test subjects were told that they would play the game twice—first as giver then as taker. In this trial, the mean offer was 45 percent and the median offer was 39 percent.

Many theoretical game theorists were skeptical of the experimental evidence and suggested that test subjects were either ill-informed or that the trials were flawed. But, as noted in chapter 1, game theory seeks to predict a player's best response, not necessarily the way a player will actually behave. Numerous experiments by behavioral economists have led researchers to conclude that givers do not behave in a purely self-interested manner, but that rational behavior was "bounded."[1] According to Camerer and Thaler (2003, p. 159), "people care how much others get."

Nash Bargaining

Nash bargaining is a two-player static game in which risk-averse rivals simultaneously bid for a share of a monetary pie of known value. If the sum of the bids is less than the size of the pie, the players receive their proposed shares. If the sum of the bids is greater than the size of the pie then each bidder receives nothing, in which case the players submit a revised bid.

Consider, for example, the following bargaining scenario involving two risk-neutral bidders. Under the terms of a collective bargaining agreement, management and labor union leaders propose the distribution of company operating profits in excess of some predetermined amount. An independent accounting firm has determined that these excess profits amount to $200 per worker. Each side submits a sealed bid for a share of these profits. To ensure fairness, both sides agree to employ the services of an independent arbiter to monitor the bargaining proceedings.

To simplify the analysis, suppose that all bids must be in $50 increments. If the sum of the bids equals $200, the players receive their bids and the game ends. If the sum of the bids is less than $200, the bidders receive their proposed distribution and the rest is donated to charity. If the sum of the bids exceeds $200, each side must resubmit their bids. To avoid the game being infinitely repeated, each side is penalized $5 per worker per round to defray monitoring costs. This static bargaining game is depicted in Fig. 18.3.

				Union		
		$0	*$50*	*$100*	*$150*	*$200*
	$0	(0, 0)	(0, 100)	(0, 120)	(0, 150)	**(0, 200)**
	$50	(50, 0)	(50, 50)	(50, 100)	**(50, 150)**	(−5, −5)
Management	*$100*	(100, 0)	(100, 50)	**(100, 100)**	(−5, −5)	(−5, −5)
	$150	(150, 0)	**(150, 50)**	(−5, −5)	(−5, −5)	(−5, −5)
	$200	**(200, 0)**	(−5, −5)	(−5, −5)	(−5, −5)	(−5, −5)

Payoffs: (Management, Union)

Fig. 18.3.

This coordination game has five Nash equilibrium strategy profiles. The strategy profile {*$150*, *$50*}, for example, is a Nash equilibrium since neither player can improve its payoff by unilaterally changing its bid. It is unlikely, however, that union representatives will be so altruistic. Similarly, the strategy profile {*$0*, *$200*} is also a Nash equilibrium, but this strategy profile is even more unlikely. It should be clear that it is in the best interest of both sides to submit bids of $100.

The game depicted in Fig. 18.3 offers clues as to why negotiations can be difficult and protracted. Ten of the twenty five strategy profiles involve payoffs that are less than the amount of excess profits available for distribution. Ten strategy profiles involve combined bids that exceed the value of the monetary pie. Fortunately, experimental and anecdotal evidence demonstrates that participants in bargaining games tend to accept an equitable split as "fair," even in the presence of multiple Nash equilibria.

Rubenstein Bargaining

We often think of bargaining as a sequential-move process. That is, an offer followed by a counteroffer until the parties come to an agreement. In this section we will discuss **Rubenstein bargaining**, which refers to sequential-move bargaining with a potentially infinite number of offers and counteroffers. While unrealistic, this assumption makes it easier to identify optimal bidding strategies in more realistic bargaining scenarios.

Before considering sequential-move bargaining games with a potentially infinite time horizon, we will begin by examining situations with a finite number of offers and counteroffers. Suppose that Beryl wants to buy Sean's car. Sean knows that Beryl is willing to pay up to $1,000; Beryl knows that Sean will not accept anything less than $500. If Beryl and Sean come to an agreement, the gain to both equals the bargaining surplus, which is the difference between the buyer's and the seller's reservation prices. In this case, the bargaining surplus is $500.

The used-car bargaining game is depicted in Fig. 18.4. To keep things simple, suppose that this bargaining game has three stages with two negotiating rounds—offer and counteroffer, followed by accept or reject. As buyer, Beryl moves first by offering Sean a price of P_B. Sean can either accept, or reject, Beryl's offer. If the offer is accepted, the game ends and the payoffs to Beryl and Sean are (\$1,000 – P_B, P_B – \$500). If the offer is rejected, Sean can make a counteroffer of P_S, where $P_S > P_B$. If the counteroffer is accepted, the payoffs to Beryl and Sean are (\$1,000 – P_S, P_S – \$500). If Beryl rejects Sean's counteroffer, the game ends and the payoffs are (0, 0).

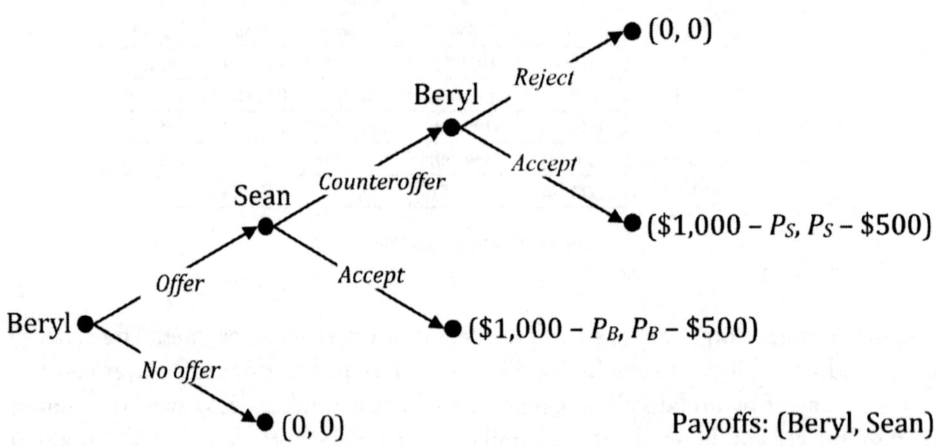

Fig. 18.4.

Suppose, for example, that Beryl offers $800. If Sean accepts Beryl's offer, the game ends and the payoffs are $1,000 – $800 = $200 and $800 – $500 = $300, respectively. The sum of the shares of the bargaining surplus is $200 + $300 = $500. Suppose that Sean does not accept Beryl's offer and makes a counteroffer of $900. If Beryl accepts Sean's counteroffer, the game ends. The payoffs to Beryl and Sean are $1,000 – $900 = $100 and $900 – $500 = $400, respectively. Once again, the sum of the shares is $100 + $400 = $500. This version of the used-car bargaining game is depicted in Fig. 18.5.

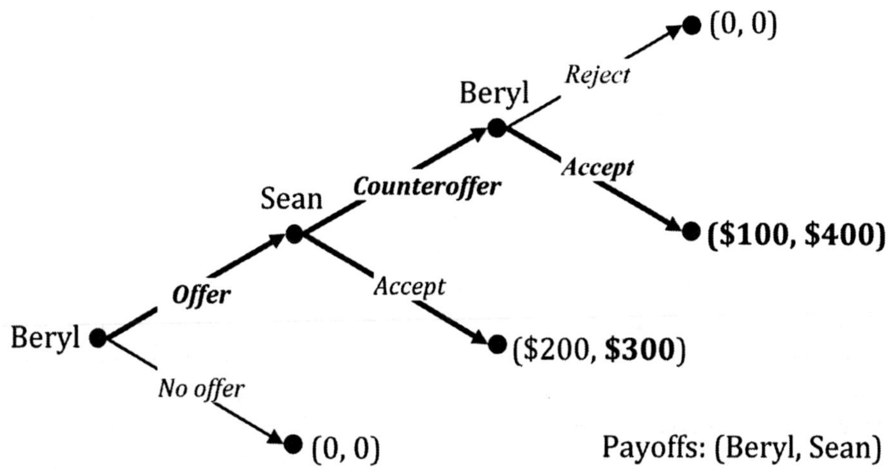

Fig. 18.5.

Using the method of backward induction, the subgame perfect equilibrium for the game in Fig. 18.5 is for Sean to reject Beryl's offer of $800 and for Beryl to accept Sean's counteroffer of $900. Beryl will receive $100 of the $500 bargaining surplus and Sean will get $400.

As it turns out, Sean's counteroffer of $900 is not optimal. Sean can do better by making an even larger counteroffer. To see why, suppose Beryl's offer remains $200, but Sean's counteroffer is $950. As before, Sean will reject Beryl's offer, and Beryl will accept Sean's counteroffer. The resulting payoffs are ($50, $450). In fact, Sean will reject any offer made by Beryl that is less than $1,000, and Beryl will accept any counteroffer made by Sean up to, but not greater than, $1,000. Beryl will make a counteroffer of $1,000 because this is how much she values Sean's car. Thus, the Nash equilibrium strategy profile for this game is {*Offer* → *Counteroffer* → *Accept*} with payoffs of (0, $500).

Last-Mover Advantage

The subgame perfect equilibrium for the used-car bargaining game with a finite number of negotiating rounds is for Sean to reject any offer that is less than his reservation price, and for Beryl to accept a counteroffer equal to Sean's reservation price. Sean's ability to extract the entire surplus of $500 regardless of Beryl's initial offer is referred to as a **last-mover advantage**.

The above example suggests that the final outcome of bargaining games depends on who makes the first offer and the number of offers and counteroffers. In the game depicted in Fig. 18.5, Beryl makes the first offer. Since there was an even number of negotiating rounds, Sean has the last-mover advantage and extracts the entire bargaining surplus. On the other hand, if there is an odd number of negotiating rounds, Beryl has the last-mover advantage and extracts the entire bargaining surplus.

Symmetric Impatience

In this section, we assume that the players incur an opportunity cost by not immediately reaching an agreement. Over time, these opportunity costs reduce the present value of the bargaining surplus. It will be demonstrated that the interest rate used to determine the present value of payoffs and the number of negotiating rounds determines the division of the bargaining surplus.

The used-car bargaining game in Fig. 18.5 involved just two negotiating rounds. Negotiations that drag on for days, weeks, months, or even years, however, may impose considerable opportunity costs on the parties involved. To see how, suppose that the used-car bargaining game involves thirty negotiating rounds with Beryl making the first offer.

Since there is an even number of negotiating rounds, Sean makes the last offer. Assume that each negotiating round takes one period, and that any delay in reaching an agreement reduces the gains to both players by $i = 0.05$, or 5 percent per period. The corresponding discount rate for each player is $\delta = 1/(1 + i) = 1/(1.05)$. Since the discount rate is the same for both players, they are said to have **symmetric impatience**. The greater the discount rate, the more "impatient" a player is to reach an agreement.

Suppose that the players are equally patient; have complete and perfect information, and have comparable negotiating skills. Each player knows the opportunity costs of failing to quickly to come to an agreement. With 30 negotiating rounds, it is impractical to illustrate the bargaining process in extensive form. For that reason, consider the information summarized in Table 18.1. Beryl is the buyer and makes the first offer in round 1. Sean is the seller and makes the last counteroffer in round 30.

Table 18.1. Symmetric Impatience

Round	*Player*	*Offer*	*Sean's Surplus*	*Beryl's Surplus*
30	Seller	$1,000.00	$500.00	$0.00
29	Buyer	976.19	476.19	23.81
28	Seller	977.32	477.32	22.68
27	Buyer	954.59	454.59	45.41
26	Seller	956.75	456.75	43.45
⋮	⋮	⋮	⋮	⋮
5	Buyer	819.93	319.93	184.07
4	Seller	824.70	324.70	175.30
3	Buyer	809.24	309.24	190.76
2	Seller	818.32	318.32	181.68
1	Buyer	803.16	303.16	196.84

Sean has the last-mover advantage. In round 30 he sells his car for $1,000.00 and extracts the entire bargaining surplus of $500. Since the present value of $500 in the twenty-ninth round is $500/1.05 = $476.19, it makes sense for Beryl to offer Sean $976.19 in the twenty-ninth round. If this offer is accepted, Beryl will realize a surplus of $500 – $476.19 = $23.81. Of course, this is not the end of the story.

Sean knows that any delay in reaching an agreement will reduce Beryl's surplus by 5 percent per round. The present value of Beryl's surplus of $23.81 in the twenty-ninth round $23.81/1.05 = $22.68 in the twenty-eighth round. Thus, Sean should make a counteroffer of $977.32. This will increase Sean's surplus to $477.32. The present value of $477.32 in the twenty-seventh round is $454.59. Thus, Beryl should make a counteroffer of $954.59, and so on. Using backward induction, Beryl's best offer in the first round is $803.16, which Sean will accept.

Even though Sean makes the last counteroffer, both players receive a share of the bargaining surplus. The shares of the surplus received by Sean and Beryl in this example are $303.16 and $196.84, respectively. This result differs from the bargaining game depicted in Fig. 18.5 in which Sean extracted the entire surplus of $500. In that game, we implicitly assumed a zero discount rate. Neither player was penalized for failing to immediately agree.

Asymmetric Impatience

Suppose that Beryl is more eager than Sean to come to an agreement. This increased anxiety is reflected in a greater opportunity cost. The more impatient is a player, the greater is that player's discount rate. This new bargaining game is characterized by **asymmetric impatience**. In this section we will show that the greater a player's relative impatience, the smaller is that player's relative share of the bargaining surplus.[2]

How will symmetric impatience affect the outcome of the used-car bargaining game? Suppose that the interest rate used to determine Beryl's interest rate is 10 percent, and the interest rate used to determine Sean's discount remains unchanged at 5 percent. The results of these negotiations are summarized in Table 18.2.

Table 18.2. Asymmetric Impatience

Round	*Player*	*Offer*	*Sean's Surplus*	*Beryl's Surplus*
30	Seller	$1,000.00	$500.00	$0.00
29	Buyer	976.19	476.19	23.81
28	Seller	978.35	478.35	21.65
27	Buyer	955.57	455.57	44.43
26	Seller	959.61	459.61	40.39
⋮	⋮	⋮	⋮	⋮
5	Buyer	849.85	349.85	150.15
4	Seller	863.50	363.50	136.50
3	Buyer	846.19	346.19	153.81
2	Seller	860.17	360.17	139.83
1	Buyer	843.02	343.02	156.98

Using backward induction, Beryl's best offer in the first round offer is $843.02, which increases Sean's share of the bargaining surplus from $303.16 to $343.02. Beryl's share of the bargaining surplus declines from $196.84 to $156.98. These results demonstrate that the relatively more patient the player, the greater will be that player's relative share of the bargaining surplus. If Beryl is more patient than Sean, Sean will still have a last-mover advantage, but will enjoy a smaller bargaining surplus.

Ariel Rubenstein (1982) demonstrated that there exists a unique subgame perfect equilibrium in sequential-move bargaining games with an infinite number of moves and countermoves.[3] Suppose that player 1 (the buyer) and player 2 (the seller) are bargaining over the division of a surplus. Player 1 makes the first offer; there is a (potentially) infinite number of negotiating rounds, and both players accept an offer when indifferent between accepting and rejecting. Player 1's discount rate is $\delta_1 = 1/(1 + i_1)$ and player 2's discount rate is $\delta_2 = 1/(1 + i_2)$. The share of the bargaining surplus going to player $2(\omega_2)$ is

$$\omega_2 = \frac{\delta_2(1-\delta_1)}{1-\delta_1\delta_2}, \tag{18.4}$$

and the share of the bargaining surplus going to player 2 (ω_2) is

$$\omega_1 = \frac{1-\delta_2}{1-\delta_1\delta_2} = 1 - \omega_2. \tag{18.5}$$

If $i_1 = 0.10$ and $i_2 = 0.5$, the share of the bargaining surplus going to player 1 is

$$\omega_2 = \frac{0.95(1-0.91)}{1-(0.95)(0.91)} = 0.631. \tag{18.6}$$

The share of the bargaining surplus going to player 2 is

$$\omega_1 = \frac{1-0.95}{1-(0.95)(0.91)} = 0.369. \tag{18.7}$$

The amounts of the surplus going to player 2 and player 1 are 0.631(\$500) = \$315.50 and 0.369(\$500) = \$184.50, respectively. Thus, player 1 should offer player 2 \$500 + \$316 = \$816 in the first round, and player 2 should accept.

Solved Exercise

Players 1 and 2 are bargaining over the price of a service contract. The highest price that player 1 is willing to pay is \$250. The lowest price that player 2 is willing to accept is \$100. Suppose that the players' interest rates are $i_1 = i_2 = 0.06$.

a. *What price should player 1 offer for the service contract in the first round? What portion of the surplus will each player receive?*
b. *Suppose that $i_1 = 0.08$ and $i_2 = 0.06$. What price should player 1 offer for the contract in the first round? What portion of the bargaining surplus will each player receive?*

Solution

a. *The bargaining surplus in this bargaining game is \$250 – \$100 = \$150. The corresponding discount rates are* $\delta_1 = \delta_2 = 1/(1.06) = 0.943$. *The fraction of this surplus going to player 2 is*

$$\omega_2 = \frac{\delta_2(1-\delta_1)}{1-\delta_1\delta_2} = \frac{0.943(1-0.943)}{1-(0.943)(0.943)} = 0.485. \tag{18.8}$$

Thus, the amount of the surplus going to player 2 is 0.485(\$150) = \$72.75. The fraction of this surplus going to player 1 is

$$\omega_1 = \frac{1-\delta_2}{1-\delta_1\delta_2} = \frac{1-0.934}{1-(0.943)(0.943)} = 0.515. \tag{18.9}$$

The amount of the surplus going to player 1 is 0.515(\$150) = \$77.25. The sum of the players' surpluses is \$72.75 + \$77.25 = \$150. Thus, player 1 should offer player 2 \$172.25 in the first round for the service contract.

b. *The discount rates for player 1 are* $\delta_1 = 1/(1.08) = 0.926$. *The fraction of this surplus going to player 1 is*

$$\omega_2 = \frac{0.943(1-0.926)}{1-(0.943)(0.926)} = 0.55. \tag{18.10}$$

Thus, the amount of the surplus going to player 2 is 0.55(\$150) = \$82.50. The fraction of the surplus going to player 1 is

$$\omega_1 = \frac{1-0.94}{1-(0.943)(0.926)} = 0.45. \tag{18.11}$$

The amount of the surplus going to player 1 is 0.45(\$150) = \$67.50. A more impatient player 1 receives a smaller share of the bargaining surplus. The sum of the players' surpluses is \$82.50 + \$67.50 = \$150. Player 1 should offer player 2 \$182.50 in the first round for the service contract and a more impatient player 2 should accept.

Practice Exercises

18.1 In bargaining games without player impatience, explain how the number of negotiating rounds and who makes the first offer affects the final outcome.

18.2 Explain what is meant by the last-mover's advantage.

18.3 What is the difference between bargaining with symmetric and asymmetric impatience? How is bargaining impatience measured?

18.4 What is the discount rate in bargaining games without impatience?

18.5 Suppose that two players are bargaining over the distribution of \$1,000 in which bids must be in increments of \$1. Each player submits a one-time bid. If the sum of the bids is less than or equal to \$1,000, each player gets the amount of their bid and the game ends. If the sum of the bids is greater than \$1,000, the game ends and the players go home empty-handed.
a. Does this game have a Nash equilibrium?
b. What is the most likely Nash equilibrium strategy profile for this game?

18.6 Two individuals are bargaining over the distribution of \$100 in which payoffs must be in increments of \$5. Each player must submit a one-time bid. If the sum of the bids is less than or equal to \$100, each player gets the amount of the bid and the game ends. If the sum of the bids is greater than \$100, the game ends and the players get nothing.
a. Does this game have a Nash equilibrium?
b. What is the most likely equilibrium strategy profile for this game?

18.7 Suppose that Bob wants to purchase a boat from Sally. Bob is willing to pay up to \$18,000, while Sally will not part with the boat for anything less than \$15,000.
a. Suppose the interest rates for Bob and Sally are $i_B = i_S = 0.05$. If there are an infinite number of bargaining rounds, what portion of the bargaining surplus will Bob and Sally receive?
b. How much should Bob offer Sally?
c. Suppose that $i_B = 0.15$ and $i_S = 0.20$. What portion of the bargaining surplus will Bob and Sally receive?
d. How much should Bob offer Sally?

18.8 Suppose that Ben wants to purchase Suzie's car. Ben is prepared to pay a maximum of \$2,000. Suzie will not accept any amount less than \$1,000.
a. Suppose that $i_B = i_S = 0.10$. If there are an infinite number of bargaining rounds, what portion of the bargaining surplus will Ben and Suzie receive?
b. How much should Ben offer Suzie?
c. Suppose that Suzie's interest rate is $i_S = 0.10$ and Ben's interest rate is $i_B = 0.20$. What portion of the bargaining surplus will Ben and Suzie receive?
d. At these interest rates, how much should Ben offer Suzie?

Notes

1. The term "bounded" rationality is generally attributed to Herbert Simon (1957) who argued that most people behave rationally and emotionally depending on the available information, cognitive limitations, and the time available to make decisions. According to Simon, bounded rationality describes the behavior of players who are limited in their ability to process information, and formulate and solve complex problems.

2. Among other things, a player with a lower discount rate may also be a more skilled negotiator.

3. The proof of this result is quite difficult. A reader-friendly explanation can be found in Osborne and Rubenstein (1994) and Binmore (1992).

PART III

GAMES WITH INCOMPLETE INFORMATION

CHAPTER NINETEEN

Decision Making and Uncertainty

Introduction

Up to this point we have assumed that players make decisions in which the payoffs were known with complete certainty. In the real world, however, decisions are frequently made under a cloud of uncertainty. Managers, for example, are uncertain as to how the public will react to a new product line, or how fluctuations in the economy, shifting consumer tastes, disruptions in resource availability, prices, labor unrest, and political instability will affect sales, revenues, and profits. Experience and market research can reduce uncertainty, but not eliminate it.

In this and the next three chapters, we will briefly review the basic elements of the **economics of information**, which involves decision making under conditions of risk and uncertainty. We begin by distinguishing between the often confused concepts of risk and uncertainty. We will then ask why a manager would choose to undertake one risky project, but not another. Alternatively, why would a manager invest in a risky project, while another manager would not? The answer often reflects their respective attitudes toward risk.

Risk and Uncertainty

The concepts of **risk** and **uncertainty** are frequently confused. Both risk and uncertainty refer to the variability of expected payoffs. In the case of risk, however, the probability distribution of alternative outcomes is known, perhaps on the basis of historical data, whereas uncertainty is subjective (Knight, 1921). We might estimate the variability of rates of return on alternative investments, but can only speculate as to who might win the next Barclay's Premier League championship. The phrase "decision making under conditions of uncertainty" is frequently used to describe the probability distribution of multiple outcomes, whether or not these probabilities have any real meaning.

The most commonly used statistical measure of risk is a data series standard deviation. When comparing alternative payoffs, a decision maker will balance expected payoffs with the volatility of payoffs over alternative states of nature. In general, a manager must be compensated with a greater expected return if a high-risk project is to be preferred over a low-risk project. How much additional compensation is required depends on the manager's attitude toward risk.

Static Games with Uncertain Payoff

Before considering the important role that risk plays in the decision making process, consider the oil-drilling game depicted in Fig. 19.1. In this game, two oil companies, PETROX and GLOMAR, must decide whether or not to drill for oil. We will initially assume that decisions are based only on expected payoffs. The riskiness of each project is assumed to play no role in the decision-making process. This is equivalent to saying that the managers are *risk neutral.*

		GLOMAR		
		D	*L*	*H*
	D	(0, 0)	(0, **132**)	(0, 104)
PETROX	*L*	(**132**, 0)	(52, 52)	(12, **70**)
	H	(104, 0)	(**70**, 12)	(**24**, **24**)

		GLOMAR		
		D	*L*	*H*
	D	(0, 0)	(**0**, –68)	(**0**, –98)
PETROX	*L*	(–68, **0**)	(–68, –68)	(–98, –98)
	H	(–98, **0**)	(–98, –98)	(–98, –98)

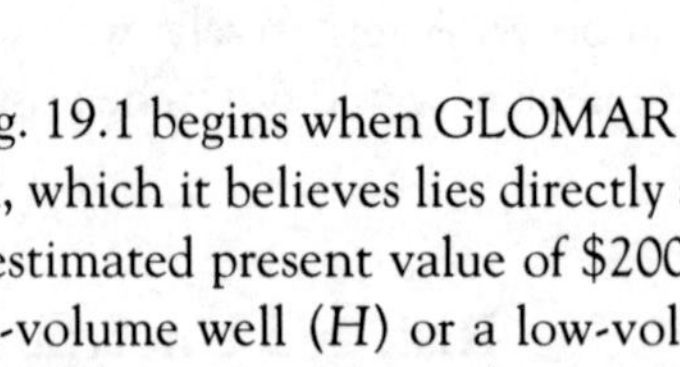

Payoffs: (PETROX, GLOMAR)

Fig. 19.1.

The static game depicted in Fig. 19.1 begins when GLOMAR purchases a two-year lease on government-owned land, which it believes lies directly above a two-million-barrel crude oil reserve with an estimated present value of $200 million. GLOMAR has the option of drilling a high-volume well (*H*) or a low-volume well (*L*). GLOMAR's rival, PETROX, has also purchased a two-year lease on adjacent land, which lies above the same crude oil deposit.

The normal-form game on the left summarizes the payoffs to each company from alternative drilling strategies if oil is present. If both companies drill the same sized wells, each will extract one-half of the total crude oil reserve. If GLOMAR and PETROX drill high-volume wells, each company will extract one million barrels in six months with a net present value of $24 million. If both companies drill low-volume wells, each will extract half the reserve with a net present value of $52 million. If both companies adopt a "don't drill" strategy (*D*), the payoff to each is $0. The remaining cells summarize the payoffs from alternative strategy profiles. The reader should verify that if oil is present, the Nash equilibrium strategy profile is {*High-volume*, *High-volume*}.

The normal-form game on the right, on the other hand, summarizes the payoffs if no oil is present. In this case, the Nash equilibrium strategy profile is {*Don't-drill*, *Don't-drill*}. The problem for both companies is that seismic testing cannot determine with certainty the presence of oil or water. On the other hand, the companies' geologists believe that there is a 60 percent chance of striking oil and a 40 percent chance of finding water (a "dry" well).

One might reasonably ask just where these probability estimates come from. In some cases, these probabilities may be estimated on the basis of empirical data. In others, they reflect the subjective opinions of stakeholders. When probability estimates do not differ or are unimportant, the resulting strategy choices are said to be robust. When probability estimates result in different strategy choices, further analysis is warranted. It may even be helpful to specify a range of outcomes that include best-case, worst-case, and most-likely scenarios.

Once probability estimates are identified, the next step is to incorporate this information into the decision-making process. One approach is to introduce a third player—conventionally referred to as "nature" (Harsanyi, 1968). Nature has no vested interest in the outcome of the game, but moves first by randomly determining states of nature with fixed probabilities that are common knowledge. Nature determines the probability of a random event at the **chance node**, which is depicted in Fig. 19.1 as an open circle.

The game depicted in Fig. 19.1 is referred to as a **static Bayesian game**, which is characterized by **incomplete information** since the players do not know for certain whether they are playing the game on the left or the right. They only know the games' probabilities. Since the managers are assumed to be risk neutral, drilling decisions are based just on expected payoffs. The expected payoffs from alternative strategy profiles are summarized in Fig. 19.2. The expected payoff from a {*High-volume*, *High-volume*} strategy profile, for example, is 0.6(24) + 0.4(–98) = –$24.8 million. The unique **Bayesian Nash equilibrium** strategy profile for this game is for both companies to drill low-volume wells.

Although expected payoffs were used to identify the pure-strategy Nash equilibrium, this should not be confused with the actual payoffs, which are realized only after the wells have been drilled. If oil is found, the payoff to each player is $52 million. If

GLOMAR

		D	*L*	*H*
	D	(0, 0)	(0, **52**)	(**0**, 23.2)
PETROX	*L*	(**52**, 0)	(**4**, **4**)	(–20, 2.8)
	H	(23.2, **0**)	(2.8, –20)	(–24.8, –24.8)

Payoffs: (PETROX, GLOMAR)

Fig. 19.2.

no oil is found, each company loses $68 million. Interestingly, if the presence of oil is known with certainty, the game depicted on the left-hand side of Fig. 19.1 is a coalition game in which the strategy profile {*Low-volume*, *Low-volume*} from cooperation is unstable in the absence of an enforcement mechanism, such as a contract.

Static Games in Extensive Form

Throughout this text, we have thus far analyzed static games using the normal form. Static games may also be illustrated using the extensive form involving a convenient analytic device called an **information set**, which is a collection of decision nodes. A player within an information set is unable to distinguish decision nodes. In Fig. 19.3, PETROX appears to move "first," which is equivalent to saying that PETROX moves without knowing how GLOMAR moves. GLOMAR also moves without knowing how PETROX moves. This is because GLOMAR cannot distinguish from among the decision nodes connected by the dashed lines.

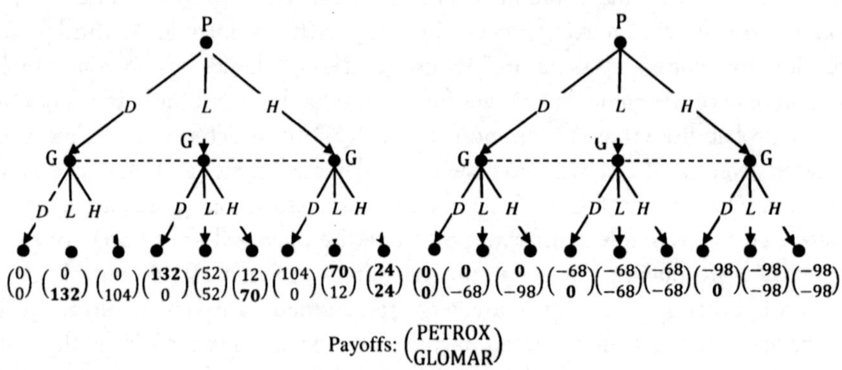

Fig. 19.3.

Since Fig. 19.3 is a static game, the search for a Nash equilibrium does not involve backward induction. Rather, it is found by examining the payoffs at the terminal nodes for each strategy profile. Consider, for example the game tree on the left-hand side of Fig. 19.3. If PETROX does not drill (*D*), GLOMAR's best response is to drill a low-volume well (*L*) with a payoff of 132. Alternatively, if GLOMAR moves high-volume, the best response by PETROX is high-volume with a payoff of 24, and so on.

We will now apply this analytic device to our search for a Nash equilibrium in Fig. 19.1. Fig. 19.4 introduces Nature who moves first by randomly determining the probability that oil is present. This game has four information sets in which the designated player cannot distinguish between decision nodes. At information set I_2, for example, GLOMAR does not know whether it is playing the game on the left or right. GLOMAR only knows the probability of being either decision node.

In Fig. 19.4 PETROX decision nodes are nested in information set I_1. PETROX does not know for certain whether oil is present. GLOMAR's six decision nodes are nested in three information sets. The expected payoffs from the nine strategy profiles in Fig. 19.2 are summarized at the terminal nodes in Fig. 19.4. For example, the payoffs for the

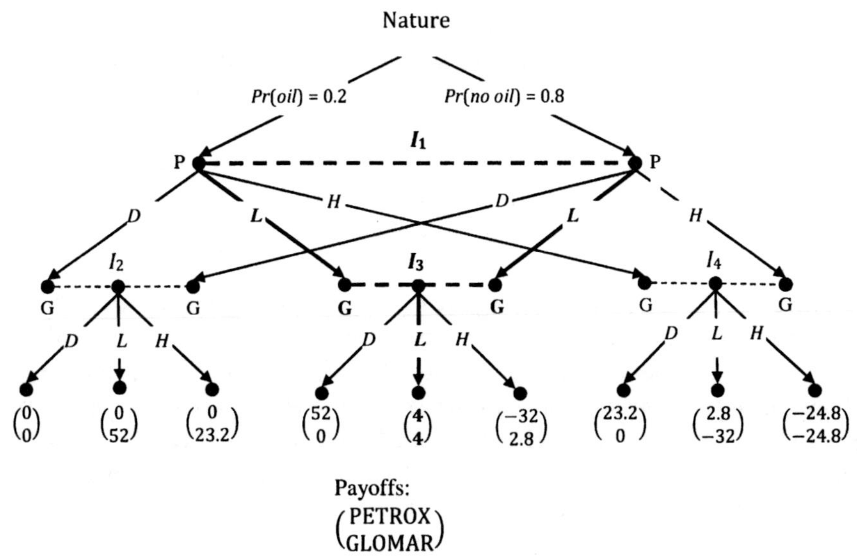

Fig. 19.4.

strategy profile {*High-volume*, *Low-volume*} are found by first tracing a *high-volume* move by PETROX from both decision nodes in information set I_1 to GLOMAR's decision nodes in information set I_4. GLOMAR's best response is low-volume, which results in payoffs of (28, −2.8). To determine the Bayesian Nash equilibrium we examine the expected payoffs for each strategy profile. The reader should verify that the Bayesian Nash equilibrium for this game is {*Low-volume*, *Low-volume*} with payoffs of (4, 4).

Dynamic Games with Uncertain Payoffs

In this section, we will search for subgame perfect equilibria in games involving risky outcomes. Consider the sequential-move complete-information pricing game in Fig. 19.5 in which A moves first. Using backward induction, the subgame perfect equilibrium is {*Low-price* → *Low-price*} with payoffs of $250,000 for both firms.

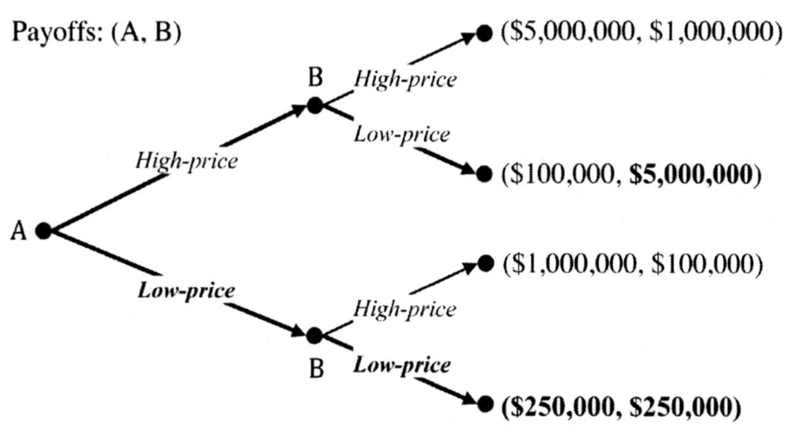

Fig. 19.5.

We will now introduce uncertainty into the decision-making process. We will continue to assume that both firms are risk neutral and that moves are based only on expected payoffs. Suppose that A believes that if it charges a high price that there is a 40 percent chance that B will charge a high price and a 60 percent chance that B will charge a low price. Similarly, if A charges a low price, there is an 80 percent chance that firm B will charge a high price and a 20 percent chance that B will charge a low price. The resulting extensive pricing game with expected payoffs is depicted in Fig. 19.6.

The first entry in the parentheses at the terminal nodes is the *expected* payoff to A. The second entry, however, is the *certain* payoff to firm B. This is because B knows how A has moved. The subgame equilibrium strategy profiles are {*High-price* → *Low-price*} and {*Low-price* → *Low-price*}. Regardless of the strategy adopted by A, B will charge a low price. A's strategy choice is based on its expected payoff. A's expected payoffs are $E(\pi_{HP})$ = 0.4($5,000,000) + 0.6($100,000) = $2,060,000 from a high-price strategy and $E(\pi_{LP})$ = 0.8($1,000,000) + 0.2($250,000) = $850,000 from a low-price strategy. The subgame perfect equilibrium, therefore, is {*High-price* → *Low-price*} since $E(\pi_{HP}) > E(\pi_{LP})$. The actual payoffs, however, are ($100,000, $5,000,000). Recall that the subgame perfect equilibrium with certain payoffs in Fig. 19.5 is {*Low-price* → *Low-price*} with payoffs of $250,000 for both firms.

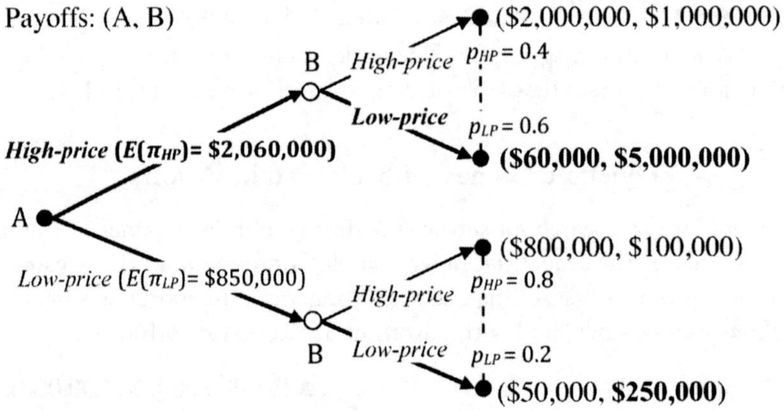

Fig. 19.6.

In the game depicted in Fig. 19.6, A was assumed to be risk neutral since the volatility of payoffs played no role. But, which strategy involves the greatest risk? A commonly used statistical measure of risk is the standard deviation.[1] The standard deviation of A's expected payoff from a *high-price* strategy is

$$\sigma_{HP} = \sqrt{[\pi_{HP} - E(\pi_{HP})]^2 p_{HP} + [\pi_{LP} - E(\pi_{HP})]^2 p_{LP}}$$
$$= \sqrt{0.4(\$2{,}940{,}000)^2 + 0.6(-\$1{,}960{,}000)^2} = \$2{,}400{,}500. \quad (19.1)$$

The standard deviation of firm A's expected payoff from a *low-price* strategy is

$$\sigma_{LP} = \sqrt{[\pi_{HP} - E(\pi_{LP})]^2 p_{HP} + [\pi_{LP} - E(\pi_{LP})]^2 p_{LP}}$$
$$= \sqrt{0.8(\$150{,}000)^2 + 0.2(-\$600{,}000)^2} = \$300{,}000. \tag{19.2}$$

Eqs. (19.1) and (19.2) reveal that a *high-price* strategy has the greater expected return, but it is also the riskiest. This strategy has an expected payoff of \$2,060,000, but the actual payoffs are \$5,000,000 or \$100,000. By contrast, a *low-price* strategy has a lower expected payoff of \$850,000, but the actual payoffs are \$1,000,000 or \$250,000, which are lower on the high side, but higher on the low side. If the volatility of returns is important, will A still adopt the riskier *high-price* strategy? The answer depends on A's attitude toward risk.

Solved Exercise

Consider the game in Fig. 19.5. Suppose that if A charges a high price then there is a 5 percent chance that B will charge a high price and a 95 percent chance that it will charge a low price. If A charges a low-price, there is a 25 percent chance that B will charge a high price and a 75 percent chance that it will charge a low price.

a. *What is the subgame perfect equilibrium if firm A is risk neutral?*
b. *Which of these projects is riskier?*

Solution

a. *This sequential-move game is depicted in Fig. 19.7. The optimal strategy for B is to charge a low price regardless of the strategy adopted by A. If A charges a high price, the expected payoff is* $E(\pi_{HP}) = 0.05(\$5{,}000{,}000) + 0.95(\$100{,}000) = \$345{,}000$.

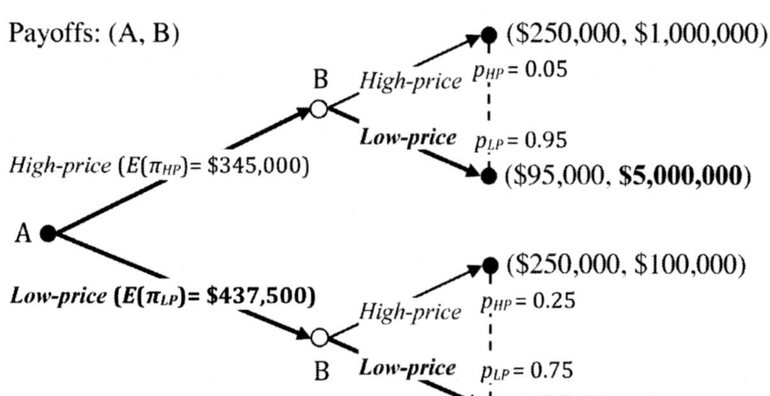

Fig. 19.7.

If A charges a low price, the expected payoff is $E(\pi_{LP}) = 0.25(\$1{,}000{,}000) + 0.75(\$250{,}000) = \$437{,}500$. *If A is risk neutral, the subgame perfect equilibrium is {Low-price → Low-price}, with actual payoffs of ($100,000, $5,000,000).*

b. *The standard deviation of firm A's payoffs from a high-price strategy is*

$$\sigma_{HP} = \sqrt{0.05(\$4{,}655{,}000)^2 + 0.95(-\$245{,}000)^2} = \$1{,}067{,}930. \quad (19.3)$$

The standard deviation of firm A's payoffs from a low-price strategy is

$$\sigma_{LP} = \sqrt{0.25(\$562{,}500)^2 + 0.75(-\$187{,}500)^2} = \$324{,}760. \quad (19.4)$$

A low-price strategy has a greater expected payoff and is also less risky. Regardless of whether A is risk neutral or risk averse, the subgame perfect equilibrium is {Low-price → Low-price} with actual payoffs of ($250,000, $250,000).

Attitudes Toward Risk

The players in the oil-drilling game depicted in Fig. 19.2 were assumed to be risk neutral. The volatility of payoffs played no role in their choice of drilling strategies. On the other hand, the decision to drill when oil is not present will result in substantial losses. Since uncertainty is characteristic of many business transactions, a firm's attitude toward risk plays an important role in the decision-making process.

Consider the situation depicted in Fig. 19.8. Player A is asked to choose between option 1 and option 2. Option 1 has a certain payoff of $950. The payoff associated with option 2 depends on the flip of a "fair" coin. If the coin comes up "heads," A receives $2,000. If the coin comes up "tails," A gets nothing. Option 2 can be described as a *lottery* with an expected payoff of 0.5($2,000) + 0.5($0) = $1,000. Which option will A choose? The answer depends on A's attitude toward risk.

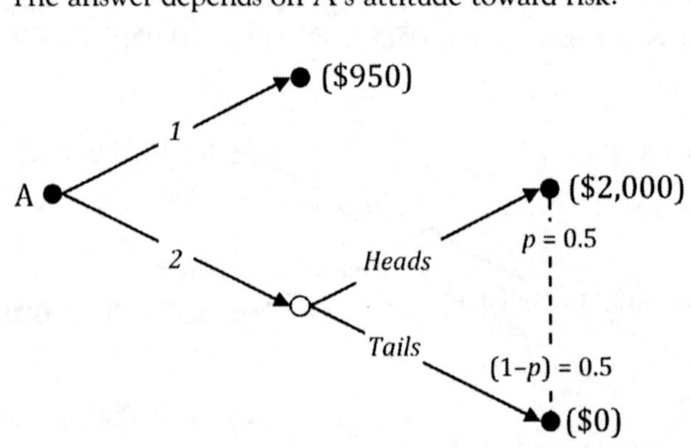

Fig. 19.8.

The choice of option 1 or option 2 depends on how much satisfaction (utility) is received by A. In general, a player's utility (U) is an increasing function of the amount of money (M) received. There is an infinite number of utility functions for which more money will generate greater utility, such as $U(M) = M$, $U(M) = M/50$, and $U(M) = 100M^{0.5}$. Since more money is preferred to less, it is safe to assume that $U(\$0) < U(\$950) < U(\$2,000)$.

Depending on A's attitude toward risk, the utility received from option 1 is $U(M) = U(\$950)$. The utility received from option 2 is $(0.5)U(\$0) + (0.5)U(\$2,000)$. Option 1 will be strictly preferred if

$$U(\$950) > (0.5)U(\$0) + (0.5)U(\$2,000). \tag{19.5}$$

Option 2 is strictly preferred if the opposite is true.

Does A prefer a certain payoff of \$950 or an expected payoff of \$1,000? To answer this question, suppose that A's utility function is $U(M) = 100M^{0.5}$, which is depicted in Fig. 19.9. Since

$$100(\$950)^{0.5} > (0.5)[100(\$0)^{0.5}] + (0.5)[100(\$2,000)^{0.5}], \tag{19.6}$$

player A prefers option 1 with a certain payoff of \$950 to option 2 with an expected payoff of \$1,000.

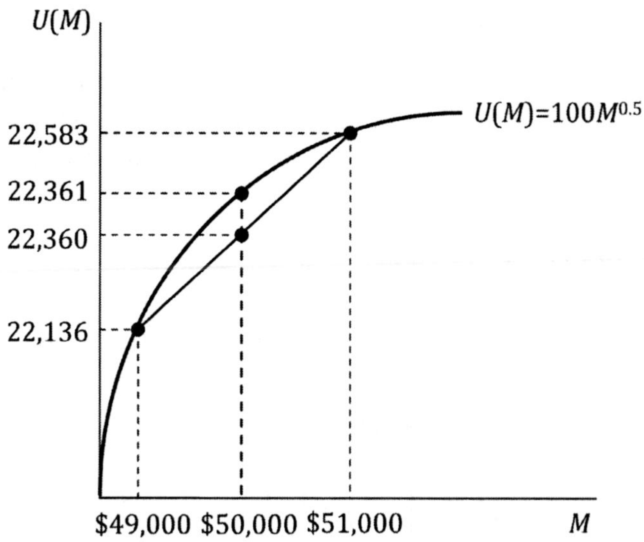

Fig. 19.9.

Risk Aversion

Risk-averse behavior was discussed by the eighteenth-century Swiss mathematician Daniel Bernoulli (1738). Bernoulli speculated that most individuals will not accept a **fair gamble** because of the diminishing **marginal utility of money**, which is declining additional satisfaction received from another dollar of money income or wealth. A fair gamble has a zero expected payoff. To illustrate Bernoulli's observation, suppose that player A is offered the following fair gamble. Player A flips a fair coin. If the coin comes up "heads," A wins \$1,000. If the coin comes up "tails," A loses \$1,000. Since there is an equal chance of a fair coin coming up heads or tails, the expected payoff is zero.

Consider Fig. 19.9. Suppose initially that A has M = \$50,000, from which she receives $U = 100(50{,}000)^{0.5} = 22{,}361$ units of satisfaction ("utils"). If A flips "heads," her total utility increases to $U = 100(51{,}000)^{0.5} = 22{,}583$, or a gain of 222 utils. If A flips "tails," her total utility declines to $U = 100(49{,}000)^{0.5} = 22{,}136$, or a loss of 225 utils. If A is risk averse, the regret from losing \$1,000 is greater than the elation from winning \$1,000. Alternatively, A's expected utility from this fair gamble is $0.5[100(49{,}000)^{0.5}] + 0.5[100(51{,}000)^{0.5}] = 22{,}359.50$ utils, which is less than the utility by refusing to play ($U(M) = 22{,}361$).

Player A is said to be **risk averse** because she prefers a sure payoff of zero by refusing to play than an expected payoff of the same amount. In general, a risk-averse individual will not accept a fair gamble since $U(M) > E[U(M)]$.[2]

How can we persuade player A to accept this fair gamble? One possibility would be to offer a *risk premium*, which is an amount of money that will make A indifferent between a certain and an expected payoff. The amount of the risk premium depends on the degree of A's risk aversion. The greater the degree of A's risk aversion, the larger the required risk premium. In Fig. 19.10, a payment of \$6.70 will make A indifferent between accepting and rejecting this fair gamble.

The more risk averse an individual, the more concave the utility function, and the greater the required risk premium. A **risk neutral** individual has a linear utility of money function, in which case a risk premium is not required to make a player

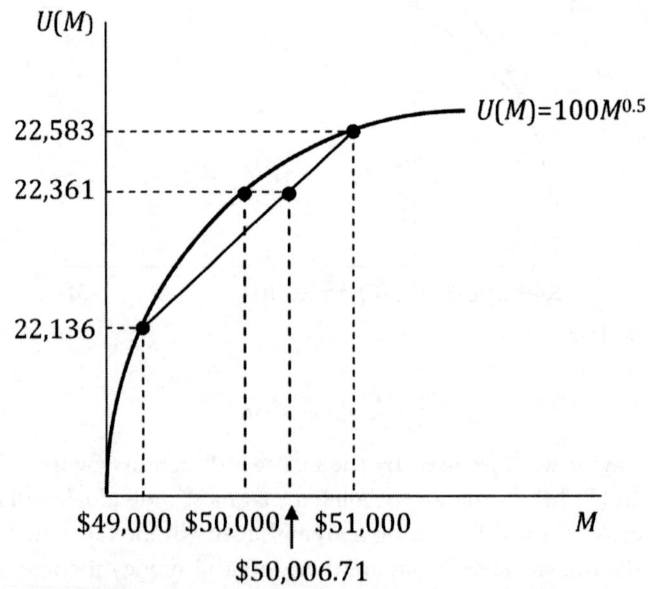

Fig. 19.10.

indifferent between a certain payoff and a risky prospect with the same expected payoff. Different degrees of risk aversion and associated risk premiums are depicted in Fig. 19.11.[3] Finally, a person who prefers the expected payoff from a risky gamble to its certainty equivalent is said to be **risk loving**. In this case, the utility function is convex (not shown).

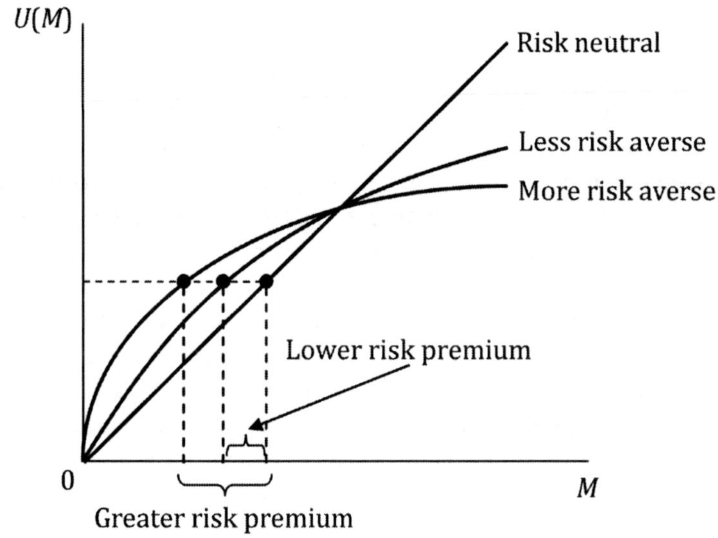

Fig. 19.11.

Solved Exercise

Adam is offered the following fair gamble. If a fair coin comes up "heads" he wins $1,000. If the coin comes up "tails" he loses $1,000. Adam's total utility of money equation is

$$U(M) = M^{1.1}. \tag{19.7}$$

Adam's initial money holdings is M = $50,000. Should Adam accept this gamble?

Solution

When M = $50,000, Adam's total utility is U = ($50,000)1.1 = 147,525.47 utils. If Adam wins $1,000, his total utility becomes U = ($51,000)1.1 = 150,774.26, for a gain of 3,248.79 utils. If Adam loses $1,000, his total utility becomes U = ($49,000)1.1 = 144,283.17, for a loss of 3,242.30 utils. The expected change in total utility by accepting the offer is

$$E(\Delta U) = (3{,}248.79)0.5 + (-3{,}241.70)0.5 = 3.55. \tag{19.8}$$

Since the expected change in utility is positive, Adam will accept this fair gamble. Alternatively, Adam's expected utility by accepting this fair gamble is E(U) = 0.5(49,000)1.1 + 0.5(51,000)1.1 = 147,528.72, which is greater than the 147,525.47 utils that

Adam receives by rejecting the offer. Adam can be described as a risk lover since he receives greater satisfaction from accepting than rejecting the offer. Adam receives positive utility from risky situations.

Understanding Risk-Averse Behavior

Understanding risk-averse behavior is important when analyzing optimal business decisions. To see why, suppose that a company is considering adding toothpaste to its product line. How can a marketing manager use his understanding of risk aversion to persuade consumers to switch to this new brand of toothpaste?

Other things being equal, consumers prefer to stay with tried-and-true brands rather than switch to new products of uncertain quality. Risk-averse consumer behavior suggests at least two possible marketing strategies. One approach would be to offer the new brand of toothpaste at a low, introductory price (risk premium) to compensate for consumer uncertainty to establish a foothold in the market. Alternatively, the firm could launch an advertising campaign designed to convince consumers the new brand of toothpaste is superior. In both instances, the objective is to raise consumers' expected net benefit from sampling the new brand.

Strategies involving product pricing, new product development, advertising, and capital investment involve an element of risk. Consider, for example, a firm that is contemplating two or more capital investment projects. Typically, managers begin by comparing the net present values for each project for the life of each investment using the equation

$$NPV = \frac{R_1 - O_1}{(1+i)^1} + \frac{R_2 - O_2}{(1+i)^2} + \cdots + \frac{R_n - O_n}{(1+i)^n} = \sum_{t=1}^{n} \frac{R_t - O_t}{(1+i)^t}, \quad (19.9)$$

where R_t are projected cash inflows (revenues), O_t are projected cash outflows (expenditures), and i is the discount rate. If these cash flows and discount rates are known with certainty, the evaluation process is relatively straightforward. Most such decisions, however, are made under a cloud of uncertainty, in which it is necessary to incorporate the manager's attitude toward risk into the evaluation process.

Fig. 19.12 depicts the risk-return tradeoffs of three managers for the same investment. Risk, which is calculated as the standard deviation (σ) of the expected returns, is measured along the horizontal axis. The expected rate of return (k) is measured along the vertical axis. The combinations of risk and return for which the managers are indifferent are called **investor indifference curves**.

Fig. 19.12 can also be used to illustrate the *risk premium*, which is the additional return required to make a risk-averse investor indifferent between a risky ($\sigma > 0$) and a risk-free ($\sigma = 0$) investment. The size of the risk premium depends on the shape of the indifference curve. Suppose, for example, that investor B is contemplating an investment with a risk of σ'. This investor would require a minimum risk premium of $(k_B - k_{rf})$, where k_{rf} is the rate of return on a risk-free investment. By contrast, the

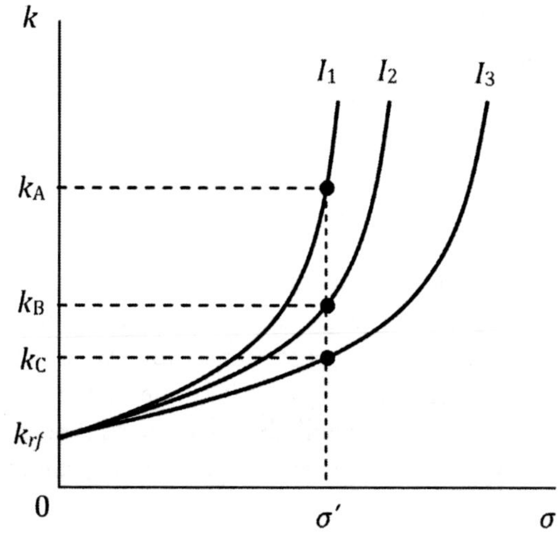

Fig. 19.12.

more risk-averse investor A requires a greater risk premium $(k_A - k_{rf}) > (k^B - k_{rf})$. Less risk-averse investor C requires a smaller risk premium $(k_C - k_{rf}) < (k_B - k_{rf})$.

An investor indifference curve may be used to evaluate **mutually exclusive** and **independent investments**. Investments are mutually exclusive if acceptance of one means rejection of all others. Investments are independent if the returns from alternative investments are unrelated and acceptance of one does not preclude other investments. Fig. 19.13 illustrates an investor's indifference curve and three mutually exclusive and equally risky investments.

In this case, a risk-averse investor will choose the investment with the greatest acceptable expected rate of return. In Fig. 19.13, only those investments in the shaded

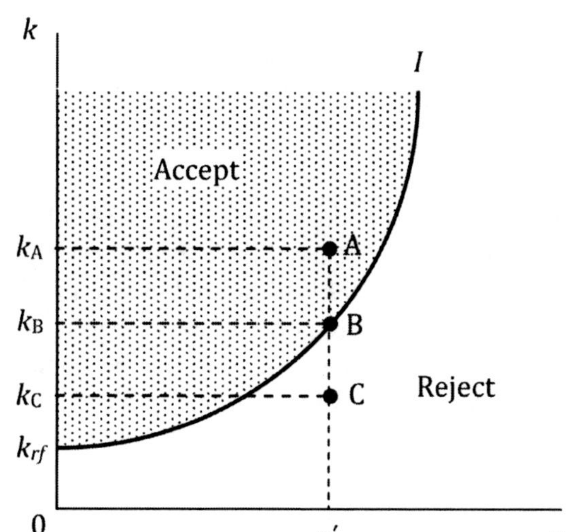

Fig. 19.13.

region will be considered. If the projects are mutually exclusive, only projects A and B will be considered. Project A will be selected since $k_A > k_B$. The rate of return k_B is the rate of return required to make this investor indifferent between investment B and a risk-free investment. Project C will be rejected since the expected rate of return (k_C) is not sufficient to compensate the investor for the risk incurred. If the three investments are independent, the investor will invest in project A and may also invest in project B, but will reject project C.

Fig. 19.14 illustrates the more complicated situation in which the expected rates of return and risk are different for each capital investment project. The expected rate of return from investment C is greater than for investment D, but it is also considerably riskier. In fact, it is so risky that it is rejected outright. If these investments are independent, this investor will choose projects B and D, and may choose A. If these investments are mutually exclusive, which project will the investor choose, and why?

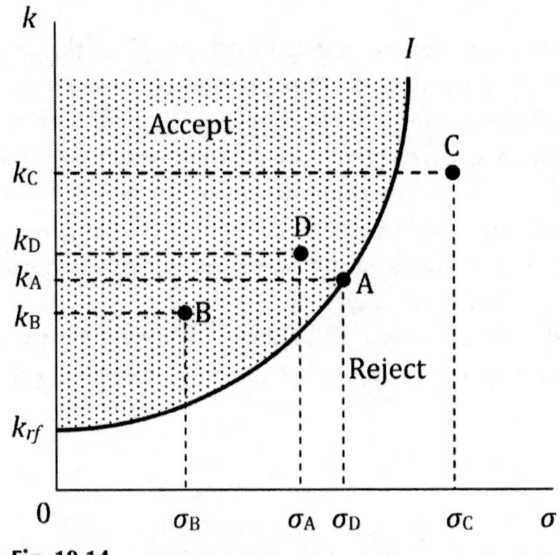

Fig. 19.14.

The risk-return indifference curve depicted in Fig. 19.14 is but one of an infinite number of such curves, with higher curves indicating a greater level of utility. A "map" of three such indifference curves is illustrated in Fig. 19.15. The lower bound of this investor indifference region corresponds to the manager's risk-free indifference curve. The most preferred investment B is located on the highest risk-return indifference curve. If the projects are mutually exclusive, project B will be accepted and project D will be rejected.

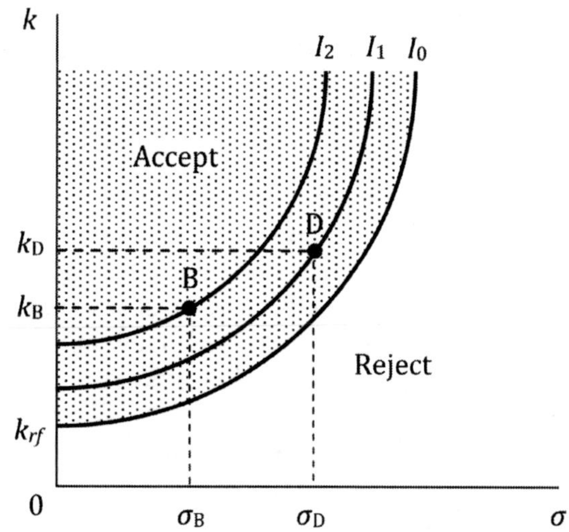

Fig. 19.15.

Practice Exercises

19.1 Explain the difference between risk and uncertainty.

19.2 Greater volatility of payoffs must be compensated by a higher expected payoff. Do you agree with this statement? For what type of person is this statement likely to be true?

19.3 What is the standard statistical measure of risk? What does it measure? If two payoffs have the same expected value, why is the payoff with the large standard deviation considered riskier than the payoff with a small standard deviation?

19.4 Can uncertainty be measured? Explain.

19.5 Explain how the marginal utility of money can be used to explain an individual's attitude toward risk.

19.6 Why do risk-averse individuals tend to reject fair gambles?

19.7 Suppose that Fred wins first prize of \$100 in a lottery at a Knights of Columbus Christmas party the same day he received a \$75 parking ticket. At the same Christmas party, Sally wins second prize of \$25. Who do you think is happier, and why?

19.8 Suppose that Sally wins first prize of \$100 in a lottery at a Knights of Columbus Christmas party the same day she loses her watch, which is valued at \$175. Fred, who also attended the Christmas party, was so distraught at having received a \$75 parking ticket, that he decided not to buy a lottery ticket. Who do you think is happier, and why?

19.9 Risk-averse individuals always reject fair gambles. Do you agree with this statement? Can you think of situations in which a risk-averse individual might accept an unfair gamble?

19.10 Why do so many risk-averse people play government-sponsored lotteries when they are unfair gambles? Would a government ever sponsor a fair lottery? Explain.

19.11 Explain the difference between a valid and an invalid information set.

19.12 Explain how a Harsanyi transformation transforms a static game with incomplete information into a static game with imperfect information. Is it possible to find a perfect Bayesian equilibrium for such static games? Why?

19.13 Rosie Hemlock offers Robin Nightshade the following wager. For a payment of $10, Rosie will pay Robin the dollar value of any card drawn from a standard 52-card deck. For example, for an ace of any suit, Rosie will pay Robin $1. For an 8 of any suit, Rosie will pay Robin $8. A ten or picture card of any suit is worth $10.

a. What is the expected value of Rosie's offer?

b. Will risk-neutral Rosie accept this offer?

19.14 Suppose that capital investment project X has an expected value of μ_X = $1,000 and a standard deviation of σ_X = $500. Suppose that capital investment project Y has an expected value of μ_Y = $1,500 and a standard deviation of σ_Y = $750. What is the relatively riskier project?

19.15 Senior management of Rubicon & Styx is trying to decide whether to advertise its world-famous hot sauce "Sergeant Garcia's Revenge" on television (campaign A) or in magazines (campaign B). The marketing department of Rubicon & Styx has estimated the probabilities of alternative sales revenues (net of advertising costs) using each of the two media outlets, which are summarized in the following table.

a. Calculate the expected revenues from each advertising campaign.

b. Calculate the standard deviation of sales revenues from each advertising campaign.

Television Revenues	*Probability*	*Magazine Revenues*	*Probability*
$5,000	0.20	$6,000	0.15
$8,000	0.30	$8,000	0.35
$11,000	0.30	$10,000	0.35
$14,000	0.20	$12,000	0.15

19.16 Using investor indifference curves, illustrate three investments in which the expected rates of return are $k_C > k_A > k_B$, and the risks associated with each project are $\sigma_C > \sigma_A > \sigma_B$. If the risk-free rate of return is k_{rf}, draw an indifference curve corresponding to each investment.

19.17 Using investor indifference curves, illustrate three investment in which the expected rates of return are $k_A > k_B > k_C$, and the risks associated with each project are $\sigma_C > \sigma_A > \sigma_B$. If the risk-free rate of return is k_{rf}, draw an indifference curve for each of these investments.

19.18 Ted Sillywalk offers Wally Wobble the fair gamble of receiving \$500 of the flip of a fair coin showing heads, and losing \$500 if the coin comes up tails. Suppose that Wally's total utility of money function is

$$U = M^{1.2}. \qquad \text{(PE19.18)}$$

a. For some positive level of money income, what is Wally's attitude toward risk?
b. If M = \$5,000, will Wally accept Ted's offer? Explain.

19.19 Consider the static oil-drilling games depicted in Figs. 19.1 and 19.2. What is the Nash equilibrium strategy profile if PETROX and GLOMAR are risk neutral and there is a 50-50 chance of finding oil?

19.20 Suppose that there is a 50-50 chance that a risk-averse individual will have an automobile accident and suffer a loss of \$100,000.

a. Calculate the cost of actuarially fair insurance if this individual's current income is \$200,000. Illustrate your answer and show that this individual prefers fair insurance to self insuring.
b. Suppose that two types of insurance policies are available. Policy 1 is fair insurance covering complete loss incurred. Policy 2 is fair insurance covering half of the loss incurred. Demonstrate that this individual will prefer policy 1 to policy 2.
c. Suppose that individuals who purchase policy 2 are likely to be more careful drivers, which reduces the cost of an accident to \$70,000. In this case, what is the cost of policy 2? Explain how some individuals might now prefer policy 2.

19.21 Sally purchases a dozen eggs. Although the cost of transporting the eggs is zero, Sally knows that there is still a 50 percent probability that she will break all of the eggs before she gets home. Knowing this, Sally considers two transportation strategies. The first strategy involves carrying all the eggs home in one trip. The second strategy involves carrying six eggs in each of two trips.

a. How many eggs are expected to survive using each strategy?
b. Suppose that Sally is risk averse. Illustrate Sally's strategy options.
c. If the cost of transporting eggs is zero, which strategy should Sally choose? Is it possible for Sally in increase her utility by increasing the number of trips? How might your answer be different if transporting eggs is costly?

19.22 A share of stock in MagnumOpus, Inc. is \$54 per share. It is equally probable that MagnumOpus shares will sell for \$50 or \$62 per share in 30 days.

a. Suppose that the strike price on a 30-day call option is \$55. What is the price of a call option if the cost of borrowing (interest rate) to finance the purchase is zero?
b. How is the price of a call option affected if the strike price increases to \$56?

c. Suppose the strike price is $56. How will the price of a call option be affected if it is equally likely that MagnumOpus shares sell for $48 or $65 in 30 days?
d. How is your answer to part c affected if the interest rate is 5 percent?
e. Consider the situation in part b. What is the price of the call option if the probability of a low price after 30 days is $p_L = 0.4$ and the probability of a high price after 30 days is $p_H = 0.6$?

Notes

1. See Appendix C.
2. The degree of curvature of the utility function can be determined by examining its first and second derivatives. If the second derivative is negative, the utility function is concave, which is characteristic of risk-averse behavior. Denoting the first and second derivatives of the utility function as $U'(M)$ and $U''(M)$, respectively, a well-known measure of risk aversion developed by Kenneth Arrow (1970) and John Pratt (1964) is $-MU''(M)/U'(M)$. Suppose, for example, that $U(M) = M^{\beta}$. The *Arrow-Pratt measure of relative risk aversion* is $-MU''(M)/U'(M) = 1 - \beta$. If $\beta = 1$, the utility function is linear and this measure of risk aversion is 0. If $0 < \beta < 1$, the utility function is concave and this measure of risk aversion is a positive fraction, which indicates risk-averse behavior.
3. Another example of the consequences of risk-averse behavior is the advantage enjoyed by chain stores and franchise operations over independently owned and operated retail outlets. A risk-averse American tourist in a foreign country prefers a first meal of predictable quality at McDonald's to unpredictable native cuisine at a local bistro. This response to product uncertainty also helps to explain why retail chain stores or franchise operations are often located in urban centers where there are a large number of out-of-town visitors.

CHAPTER TWENTY

Adverse Selection

Introduction

There are many problems in economics and business where some market participants have more or better information than others about the transaction of a good or service. This problem of **asymmetric information** leads to problems of **adverse selection** and *moral hazard*. Adverse selection is an *ex ante* problem in that it arises before the transaction takes place. An often cited example of this arises in the loan market. Borrowers know more about their own creditworthiness than do lenders. Since lenders cannot distinguish good credit risks from bad credit risks, the interest rate charged reflects the average borrowers' credit worthiness. Bad credit risks will happily pay the lower (for them) interest rate, while good credit risks will drop out of the market. As the market becomes crowded with bad credit risks, lenders may decide not to lend at all, even though there are good credit risks who want to borrow.

Moral hazard is an *ex post* problem in that it arises after the transaction has taken place. **Moral hazard** is the risk (hazard) that one party to a transaction behaves in an undesirable (immoral) manner. Suppose, for example, that the owner of a company obtains a bank loan to invest in new machinery. The bank's approval was based on the belief that the capital investment will generate a rate of return sufficient to nominally service the debt. Once the loan is made, however, the borrower has an incentive to use the loan for some other, riskier, purpose that promises a higher rate of return. Because moral hazard lowers the probability that the loan will be repaid, the bank may decide not to lend without adequate safeguards. In loan contracts, these safeguards take the form of *restrictive covenants*.

The Market for Lemons

Nobel laureate George Akerlof discussed the problem associated with adverse selection in the used-car market. Professor Akerlof (1970) began by identifying two types of used cars: good used cars (peaches) and bad used cars (lemons). In this market, there are no used-car dealers. Transactions directly involve buyers and sellers. The problem of asymmetric information arises because sellers know the value of their used cars, but buyers do not.

Buyers will pay a low price for low-quality used cars (lemons) and a high price for high-quality used cars (peaches). Since they are unable to distinguish lemons from peaches, the market price reflects the average quality of used cars. Owners of lemons are happy to receive the higher (for them) average price. Owners of peaches will not want to sell at the lower (for them) average price and will drop out of the used-car market. As a result, the market will become crowded with lemons. Buyers of used cars know this, so few used cars will be purchased. If the problem of asymmetric information is not resolved, the used-car market could cease to exist.

To illustrate the problem of adverse selection in the used-car market, consider the used-car bargaining game depicted in Fig. 20.1, which is characterized by incomplete information. We will assume that the players are risk neutral and that a dollar yields equivalent utility for both players.

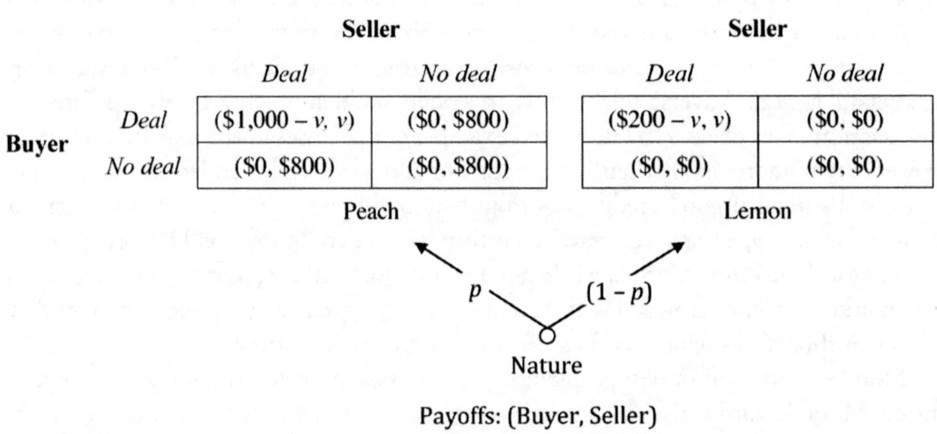

Fig. 20.1.

In this bargaining game, the buyer values a peach at v_B = \$1,000 and a lemon at v_B = \$200. The seller values a peach at v_S = \$800 and a lemon at v_S = \$0. The market price (v) reflects average value of the seller's used car. This is a problem of asymmetric information since only the seller knows for certain whether her used car is a peach or a lemon. In this game, Nature moves first and determines the probability that the used car is a peach (p) or a lemon ($1 - p$). In this one-time, static, competitive game, the strategy choices are *deal* and *not deal.*

In this game, the buyer will want to deal if his expected bargaining surplus is non-negative, that is

$$p(\$1{,}000 - v) + (1 - p)(\$200 - v) \geq 0, \tag{20.1}$$

which simplifies to $p \geq (v - \$200)/\$800$. Condition (20.1) says that the buyer will pay the market price if the probability that the used car is a peach is sufficiently positive. Suppose that an Internet search reveals that the *Kelley Blue Book* value of the seller's make and model is $v = \$900$.[1] Substituting this value into condition (20.1) we get $p \geq (\$900 - \$200)/\$800 = 0.875$. This says that the buyer will pay the market price if he believes there is at least an 87.5 percent chance that the used car is a peach.

The lemons problem has multiple applications. In the primary market for financial securities, for example, sellers of stocks and bonds have better information about a company's financial strengths or weaknesses than do buyers. Buyers of equities (stocks) are less able to distinguish high-earnings, low-risk companies from low-earnings, high-risk companies than are the issuing firms. Thus, equities prices reflect the average value of these companies. Sellers of low-quality stocks are happy to sell at the (higher) average price. Sellers of high-quality stocks will not want to sell at the (lower) average price. Similarly, buyers of debt (bonds) are less able to distinguish high-default-risk companies from low-default-risk companies than are the bond issuers. Thus, interest rates will tend to reflect the average of default risk. Sellers of high-default-risk bonds are happy to borrow at the (lower) average interest rate, while sellers of low-default-risk bonds will not want to borrow at the (higher) average interest rate.

As in the used-car market, unless the problem of asymmetric information is resolved, adverse selection will result in the market becoming crowded with low quality financial securities. The result will be inefficient equity and bond markets. If the problem of adverse selection becomes severe enough, the securities markets could disappear entirely. Of course, used-car, equity, and bond markets do exist, which means remedies have been devised to overcome adverse selection problems arising from asymmetric information.[2]

Solved Exercise

Consider again the lemons problem in Fig. 20.1.

a. *Suppose the* Kelley Blue Book *value is \$920. What is the minimum probability that the used car is a peach for the buyer to pay this price?*
b. *What is the* Kelley Blue Book *value if the probability that the used car is a peach is* $p = 0.6$*?*

Solution

a. *Substituting the* Kelley Blue Book *value into condition (20.1) we obtain*

$$p \geq \frac{v - \$200}{\$800} = \frac{\$900 - \$200}{\$800} = 0.9. \tag{20.2}$$

The buyer must believe that there is at least a 90 percent probability that the used car is a peach before paying $920 for this used car.

b. The Kelley Blue Book *value may be found by solving*

$$0.6 \geq \frac{v - \$200}{\$800}. \quad (20.3)$$

Solving Eq. (20.3), the Kelley Blue Book *value must be at least v = $680.*

Practice Exercises

20.1 Asymmetric information gives rise to the problems of adverse selection and moral hazard. Explain.

20.2 What is the market for lemons? In what way do you think used-car dealers help reduce the problem of adverse selection?

20.3 How does the lemons problem manifest itself in the labor market?

20.4 How might government regulation help remedy the problem of asymmetric information in financial markets?

Notes

1. The *Kelley Blue Book* is the largest vehicle valuation company in the United States. Based in Irvine, California, the company's website is considered the most reliable source for establishing the average market price of new and used cars.

2. A solution to adverse selection in financial markets is to increase the flow of information. Financial services companies such as Standard and Poor's, Moody's, and Value Line collect and analyze financial and economic data, which they sell to investors in financial securities. The private production of information, however, does not completely solve the problem of adverse selection because of the free-rider problem. This occurs when investors who have not paid for private information mimic the purchases of those who have. This drives up the prices of the financial securities, which reduces the incentive to purchase information. The result is that less private information is produced.

CHAPTER TWENTY-ONE

Incentive Contracts

Introduction

A firm is an activity that transforms scarce factors of production into outputs of goods and services for sale in the market. The manner in which productive resources are combined and organized depends on the objectives of the firm's owners, managers, or both. In some cases, the owner and manager are the same. In the case of publicly-owned companies, decisions are made by managers who are the designated agents of shareholders.

Although there are many possible organizational objectives, economists usually assume that the firm is a profit maximizer. This behavioral assumption is central to the neoclassical theory of the firm, which posits the firm is a profit maximizing "black box" that transforms inputs into outputs. While the precise contents are unknown, it contains the "secret formula" that gives the firm its competitive advantage.[1]

Although the assumption of profit maximization is intuitively appealing, the objective of most publicly-held corporations is to maximize shareholder value, which is equivalent to maximizing the net present value of future earnings. This requires that the firm procures inputs at the least cost, minimizes transactions costs, and that nonhuman resources are efficiently employed and human resources (managers and workers) put forth their best efforts. Unfortunately, the organizational structure of many large corporations gives rise to the **principal-agent problem**, which occurs when one group of stakeholders called agents fails to represent the best interests of another group of stakeholders called principals because their interests are not properly aligned.

Principal-Agent Problem

A distinguishing feature of many large corporations is that they are not owner-operated. Responsibility for the day-to-day operations of the firm is delegated to managers who are supposed to represent the best interests of shareholders. But how can absentee owners be assured that managers are doing everything they can to maximize shareholder value?

Classical theories of economic behavior are based on the assumption of rational self-interest. Individuals make choices by comparing the marginal benefits and marginal costs of their decisions. A fundamental incentive problem may emerge when a manager's compensation is not tied to a company's performance. If a company underperforms, was it because of circumstances beyond managers' control, or was it because the managers put their self-interests before those of the owners? Because the shareholders are absentee owners, it can be difficult to determine the cause.

Suppose, for example, that a manager is paid a lucrative fixed annual salary regardless of effort. Unless properly motivated, a fundamental conflict may arise between owners who value profit, and managers who derive utility from the non-pecuniary perquisites that come with the job. Fixed compensation schemes will not provide managers with an incentive to put forth their best efforts on behalf of shareholders. In the absence of constant monitoring, managers whose compensation is fixed have an incentive to engage in activities that are in their own best interest, but could adversely affect the company's bottom line.

Incentive Contracts

To overcome the owner-manager principal-agent problem it is necessary to align the interests of owners and managers. One way to accomplish this is with a properly structured **incentive contract**, which may include provisions for *profit-sharing*, *stock options*, *performance bonuses*, and other rewards that link a manager's compensation to the company's performance.

Suppose, for example, that a manager is offered a share of the company's profits in addition to a modest base salary. When gross compensation is tied to the company's performance, the manager may have an incentive to put forth his or her best effort on behalf of shareholders. If this increased effort results in greater profit, both the owners and managers benefit. Of course, the extra effort put forth by the manager depends on the marginal utility of an additional dollar of income. Owners must take care not to offer a base salary that is too generous; otherwise the added income may not provide the intended incentive.

Principal-Agent Problem with Moral Hazard

Incentive contracts can be an effective means for ensuring that agents act in the best interest of principals, but only if both parties have complete information. Since owners cannot directly observe the behavior of managers, however, a fundamental problem of asymmetric information arises. Because of this, it is not possible to structure and enforce an incentive contract based on unobservable effort. It is only possible to structure a contract based on observable performance. This situation is referred to as the *manager-owner principal-agent problem with moral hazard.* Unfortunately, an incentive contract that links compensation to performance may raise other problems. Suppose, for example, that a manager is in a position to engage in unethical or illegal activities that temporarily raise the company's share prices at the expense of future earnings.

To highlight the issues involved when structuring an incentive contract in the presence of asymmetric information, consider the principal-agent problem with moral hazard game depicted in Fig. 21.1.[2] In this game, the owner (O) of a company must decide whether to offer a contract to a new project manager (M). The success of the project depends on whether the manager is diligent (*D*) or negligent (*N*) and a random factor, such as the state of the economy. If the project succeeds (*S*), the owner's profit is π_2. If the project fails (*F*), the owner's profit will be a lesser π_1, where $\pi_1 < \pi_2$.

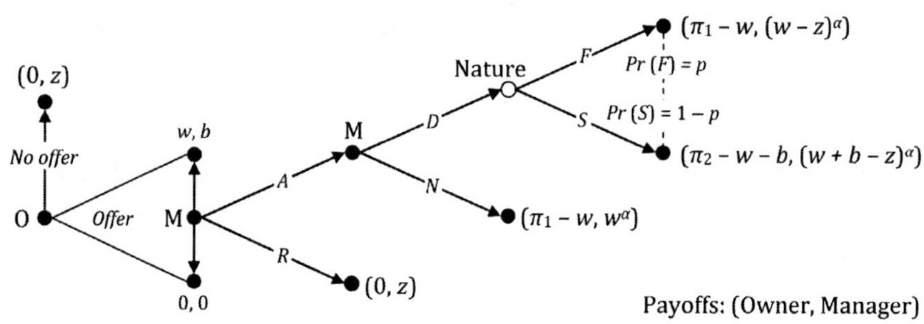

Fig. 21.1.

The principal-agent game begins when the owner considers offering a contract to a new project manager. A contract offer consists of a fixed wage (w) that is paid regardless of the project's success or failure and a performance bonus (b) if the project is a success. The vertical double arrow indicates that there are an infinite number of contracts corresponding to different wage-bonus combinations that the owner can offer. The objective of the owner is to identify the lowest-cost incentive contract that will promote diligence that is accepted by the project manager.

If no offer is made, the game ends and the owner's profit is zero and the manager receives positive utility from an amount z from the next best alternative activity. These are the same payoffs if the contract offer is rejected. If the contract is accepted (*A*), the project manager decides to be diligent or negligent. Diligence does not guarantee that the project will succeed, which also depends on the state of the economy. By contrast, negligence ensures that the project will fail. Assume that the manager's utility function is U (M) = M α = $(w + b)\alpha$, and that the project manager is risk averse ($0 < \alpha < 1$). We also assume that the owner is risk neutral by pooling the risks of multiple projects.

While negligence ensures failure, diligence does not guarantee success. The principal-agent game depicted in Fig. 21.1 is characterized by incomplete information in which Nature determines the probability p of failure or success ($1 - p$). The payoffs if the project fails are $(\pi_1 - w, (w - z)\alpha)$. The payoff to the owner if the project fails is the same as the payoff when the manager is negligent. The payoff to the manager is the utility received from the guaranteed wage less the value of the foregone activity by rejecting the contract. If the project succeeds, the payoffs are $(\pi_2 - w - b, (w + b - z)\alpha)$. The owner receives π_2 less the manager's wage and performance bonus. The payoff to the manager is the utility received from guaranteed wage and performance bonus less value of the foregone activity.

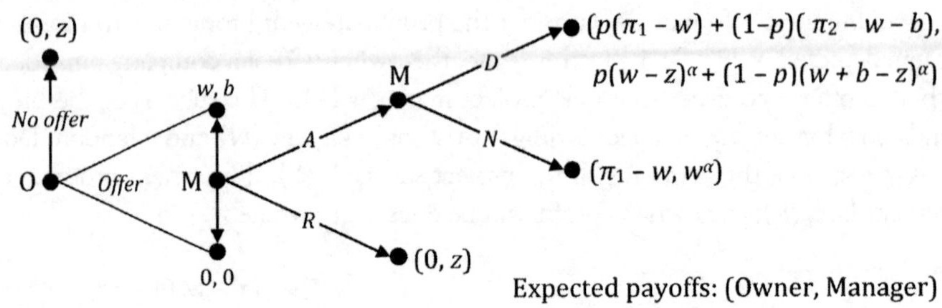

Fig. 21.2.

The principal-agent game with expected payoffs is depicted in Fig. 21.2. Bayesian subgame perfection is determined through backward induction. Proceeding left from the terminal nodes, the manager will be diligent if the expected payoff is greater than the payoff from negligence, that is

$$p(w - z)^{\alpha} + (1-p)(w + b - z)^{\alpha} \geq w^{\alpha}. \tag{21.1}$$

We then move to the immediately preceding decision node and ask whether the manager should accept or reject the contract. The manager should accept if the expected utility from the payoff is at least as great as the value of the alternative activity by rejecting the contract, that is

$$p(w - z)^{\alpha} + (1-p)(w + b - z)^{\alpha} \geq z. \tag{21.2}$$

Meanwhile, it will be in the owner's best interest for the manager to accept the contract if the expected profit less the manager's compensation from being diligent is greater than the payoff from negligence, that is

$$p(\pi_1 - w) + (1-p)(\pi_2 - w - b) \geq \pi_1 - w. \tag{21.3}$$

Inequality (21.3) may be simplified to yield $\pi_2 \geq \pi_1 + b$. That is, for an incentive contract to make sense, the owner's expected net profit when the manager is diligent must be at least as great as when the manager is negligent.

What wage and bonus package should the owner offer? It is in the owner's best interest to pay the minimum amount necessary to get the manager to put forth maximum effort. After setting (21.1) and (21.2) to equalities and solving we obtain $w\alpha = z$. Substituting this amount into the transformed inequality (21.2) we get

$$p(w - z)^{\alpha} + (1 - p)(w + b - z)^{\alpha} = z. \tag{21.4}$$

Given the probability distribution of the project's success or failure, the manager's degree of risk aversion, and the value of the manager's alternative activity, Eq. (21.4) may be solved to obtain the minimum bonus required for the manager to accept the contract. Once the optimal bonus is identified, the equation from (21.3) can be used to determine if it is in the firm's best interest to offer the manager a contract. Finally, the risk premium required to make the manager indifferent between diligence and negligence is the difference between the expected payoff from risk neutrality ($\alpha = 1$) and risk aversion ($0 < \alpha < 1$).

Solved Exercise

Consider the above example in which the firm's profit if the project fails is $\pi_1 = 0$ *and* $\pi_2 = 100$ *if the project is a success. Suppose that there is a 50-50 chance that the project will succeed and that the manager's degree of risk aversion is* $\alpha = 0.75$.

a. *What is the owner's optimal wage that the owner should offer if the value of the manager's alternative activity is 10?*
b. *What is the owner's optimal bonus offer?*
c. *What is the subgame-perfect Bayesian equilibrium?*
d. *What risk premium is required to make the manager indifferent between negligence and diligence?*

Solution

a. *The optimal wage can be found by solving* $w\alpha = z$. *Substituting the information provided, the owner should offer* $w = 21.54$.
b. *Substituting the values provided and the optimal wage into Eq. (21.4), the optimal bonus is marginally greater than* $b = 21.37$.
c. *The principal-agent game with expected payoffs is depicted in Fig. 21.3. Bayesian subgame perfection is determined through backward induction. In this principal-owner game, the owner will offer a contract, which the manager will accept and work diligently for the project's success, that is, {Offer → Accept → Diligent}.*

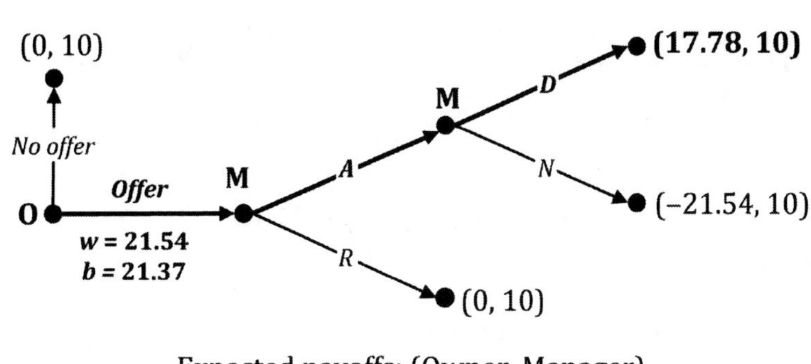

Fig. 21.3.

d. *The expected payoff from risk neutrality is*

$$0.5(21.54 - 10)^1 + 0.5(21.54 - 10 + 21.37)^1 = 22.23. \quad (21.5)$$

The expected payoff from risk aversion is

$$0.5(21.54 - 10)^{0.75} + 0.5(21.54 - 10 + 21.37)^{0.75} = 10. \quad (21.6)$$

The risk premium that is required to make the manager indifferent between negligence and diligence is 22.23 – 10 = 12.23.

Solved Exercise

Suppose that in the above solved exercise that the manager's degree of risk aversion is α = 0.5.

a. *What is the owner's optimal wage offer?*
b. *What is the owner's optimal bonus offer?*
c. *What is the subgame-perfect Bayesian equilibrium?*
d. *What risk premium is the manager's risk premium?*

Solution

a. *Substituting the information provided into* $w = z$, *the owner's optimal wage offer* $w = 100$.
b. *From Eq. (21.4), the owner's optimal bonus offer is something marginally greater than* $b = 20.67$.
c. *The revised principal-agent game with expected payoffs is depicted in Fig. 21.4. Since the owner will suffer an expected loss of 60.33 the owner will not undertake the project. The subgame-perfect Bayesian equilibrium is for the owner not to make an offer {No offer}.*

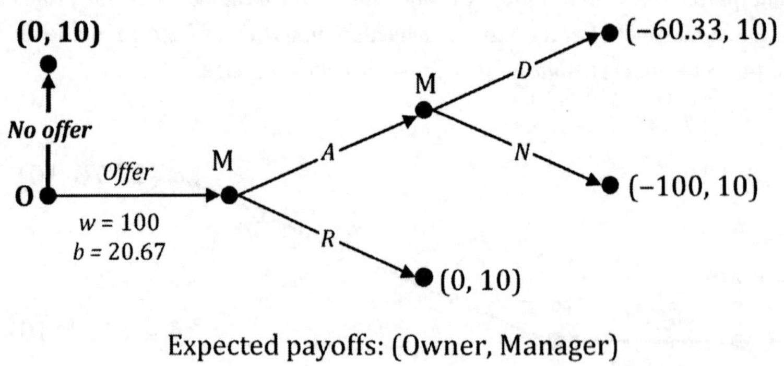

Fig. 21.4.

d. *The expected payoff from risk neutrality is*

$$0.5(100 - 10)^1 + 0.5(100 - 10 + 20.67)^1 = 100.34. \quad (21.7)$$

The expected payoff from risk aversion is

$$0.5(100 - 10)^{0.5} + 0.5(100 - 10 + 20.67)^{0.5} = 10. \quad (21.8)$$

The risk premium is 100.34 – 10 = 90.34. The greater risk premium reflects the significant increase in the manager's risk aversion.

Practice Exercises

21.1 One solution to the principal-agent problem in restaurants is the tips earned by waiters and waitresses in addition to a small base salary. Discuss a potential problem for management with this type of revenue-based incentive scheme.

21.2 Employees of fast-food restaurants do not earn tips as do waiters and waitresses in restaurants. Discuss possible solutions to the manager-worker principal-agent problem in fast-food restaurants.

21.3 Explain why frequent spot checks by managers to encourage workers to put forth their best effort may not always be in the best interest of the firm's owners.

21.4 Under what condition is the assumption of profit maximization equivalent to shareholder-wealth maximization?

21.5 Refer to the practice exercise in this chapter. Suppose that there is a 75 percent chance that the project will fail.
   a. What is the owner's optimal wage offer?
   b. What is the owner's optimal bonus offer?
   c. What is the subgame-perfect Bayesian equilibrium?

21.6 Based on the principal-agent model presented in this chapter, what effect will an increase in the project's probability of success have on the owner's optimal wage and bonus offers?

21.7 Based on the principal-agent model presented in this chapter, what affect will an increase in the manager's degree of risk aversion have on the owner's optimal wage and bonus offers?

Notes

1. Neoclassical theory makes no attempt to explain what actually goes on inside this "black box," although the underlying production function is assumed to exhibit desirable mathematical properties, such as the law of diminishing returns, returns to scale, and substitutability between and among productive inputs. The appeal of the neoclassical model is its application to a wide range of profit-maximizing firms and market situations.

2. The example in this chapter was motivated by Grossman and Hart (1983).

PART IV

GAMES WITH IMPERFECT INFORMATION

CHAPTER TWENTY-TWO

Information Sets

Introduction

The interesting thing about many strategic situations is that rivals often possess different information at various stages of the game. We have already seen that many games are characterized by incomplete information in that the payoffs from alternative strategies are not known with certainty. This uncertainty was handled analytically by applying a Harsanyi transformation in which Nature moves first by randomly determining states of nature. The resulting probability distribution was assumed to be common knowledge. In this and the next two chapters, we will explore situations in which private information is not equally shared by all players. Such games are said to be characterized by **imperfect information.**[1]

In many cases, private information has something to do with the different attributes and characteristics of players. Imperfect information results when Nature moves first by determining the probability distribution of these **player types**. Individual players privately observe their own type, which defines their actions, but do not know their rivals' types. This problem of asymmetric information can arise, for example, when a firm wants to hire a new worker. While the applicant knows with certainty whether he or she is a high-productivity or a low-productivity type, the firm does not. One way, but not the only way, to deal with this problem is for the players to form conditional expectations about payoffs, and to update these expectations as more information becomes available—a procedure known as *Bayesian updating*.

Information Sets

When analyzing dynamic games in which the players possess private information it is essential that we keep track of what the players know and when. Recall from chapter

19 that an *information set* is a collection of decision nodes. A player within an information set is unable to distinguish decision nodes. Information sets makes it possible to keep track of a player's different decision nodes that appear the same.

Consider, for example, the game depicted in Fig. 22.1. In this game, there are three decision nodes. A's decision node is at the root of the game tree. B has two decision nodes. Unlike dynamic games with complete and perfect information, however, B is not able to distinguish between decision nodes. B's decision nodes constitute an information set, which is designated by a dashed line that connects them. The collection of all information sets constitutes an **information partition**.

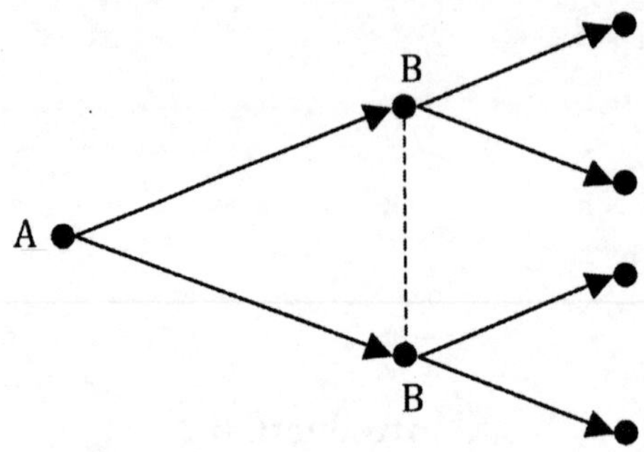

Fig. 22.1.

In Fig. 22.1, $\{B_1, B_2\}$ is called a **valid information set** because player B cannot distinguish between these decision nodes. There are three conditions that must be satisfied for an information set to be valid. First, all decision nodes in the set must belong to the same player. Second, no decision node in the set may precede any other decision node in the same set. Finally, a player must make the same move from any decision node within the set.

An example of an **invalid information set** is depicted in Fig. 22.2 because it contains the decision nodes of two different players (B and C). Player B knows her location in the game tree because player C is located at the only other possible decision node. Fig. 22.3 depicts two invalid information sets $\{A, B_1\}$ and $\{A, B_2\}$. In each case A knows where he is located because there are no prior moves. B knows where she is located because player A has already moved. Knowing A's move makes it possible for B to know her location.

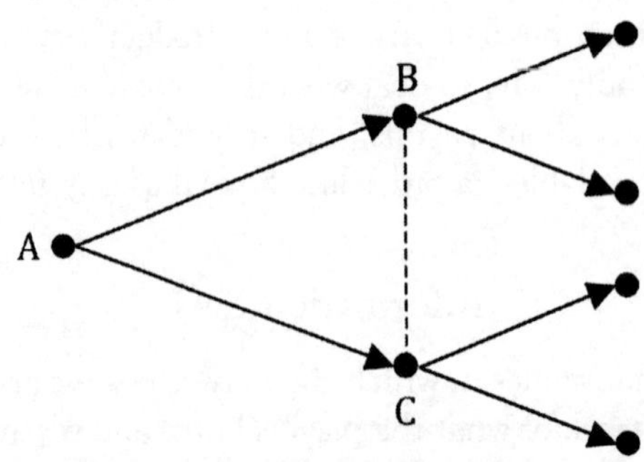

Fig. 22.2.

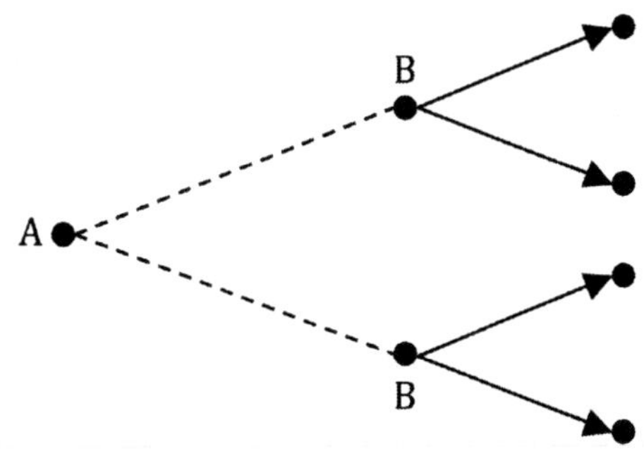

Fig. 22.3.

Valid information sets are not new. Static games involve valid information sets. Recall that in a static game the players move without knowing the strategy choices of rivals until all moves have been made. Static games are characterized by complete, but imperfect, information. In Fig. 2.1, for example, the prisoner's dilemma was depicted in extensive form using a valid information set. Suspect A appears to move "first," which is equivalent to saying that A moves without knowing B's move. By nesting B's decision nodes in an information set, B does not know whether she is located at the upper or lower decision node. In other words, B also moves without knowing how her rival has moved until all moves have been made.

The search for a Nash equilibrium in Fig. 2.1 did not involve backward induction. Since neither player knows the other's move until the end of the game, it is found by directly examining the payoffs at the terminal nodes for each strategy profile. If A moves *silent*, B will *confess*. If A moves *confess*, suspect B will *confess*. *Confess* is a strictly-dominant strategy for B. Since the payoffs are symmetrical, confess is a strictly-dominant strategy for A as well.

We can also use Fig. 2.1 to distinguish between a proper and an **improper subgame**. Recall that a proper subgame begins at subroots in sequential-move games with perfect information. The subgames beginning at either of B's decision nodes in Fig. 2.1 are not proper because they belong to a valid information set. B does not know which subgame she is playing. For dynamic games with incomplete information, a proper subgame can only begin at subroots that are not included within a valid information set. An improper subgame is any subgame with a subroot located within a valid information set.

The validity of information sets also depends on the assumption of **total (perfect) recall**. Sequential-move games with perfect information assume that players recall all prior moves, including moves made by other players. The implications of total recall are depicted in Fig. 22.4.

Consider the information sets $\{A_1\}$, $\{A_2\}$, and $\{A_3, A_4\}$. All of these decision nodes belong to player A, who is seemingly able to distinguish between A_1 and A_2, but who is not able to distinguish between A_3 and A_4. But this cannot be the case since A must move *right* to get to A_3. In a similar manner, A must move *left* to get to A_4. For A_3 and A_4 to constitute a valid information set, A must have forgotten his prior move, which is ruled out by the assumption of total recall. Thus, the information set $\{A_3, A_4\}$ is invalid.

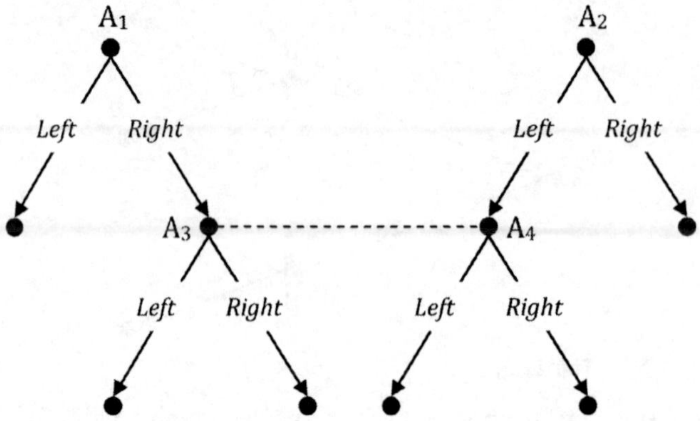

Fig. 22.4.

The assumption of total recall also rules out the situation in Fig. 22.5. In this game, information set {B_3, B_4} must be invalid. The reason for this is not difficult to see. Since players are assumed to remember everything, including the moves made by other players, the only way for B to reach B_3 is for A to move *left*. Similarly, the only way for B to reach B_4 is for A to move *right*. {B_3, B_4} can only be a valid information set if B were to forget how A moved.

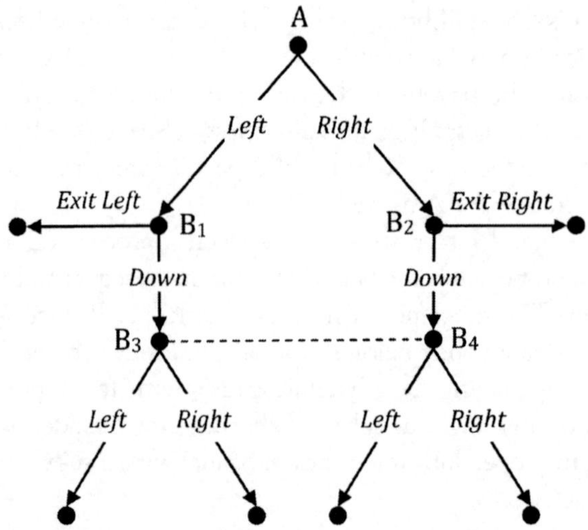

Fig. 22.5.

Bayesian Updating

In dynamic games with imperfect (asymmetric) information, a player's prior moves may reveal private information to other players, who will incorporate this information into their own decisions. The process by which players revise prior probability estimates on the basis of new information is referred to as **Bayesian updating.**

To illustrate the process of Bayesian updating, consider the following situation in which GLOMAR has purchased a two-year lease to drill on government land. The company believes (but does not know) that the land lies directly above a four-million barrel crude oil deposit. It is generally believed that there is a 20 percent chance that oil is present. Because the probability of finding oil is low, high development and extraction costs make drilling infeasible. To reduce risk, GLOMAR conducts its own seismic testing, which is not foolproof since it can produce a *false positive* (i.e., that oil exists when it does not) or a *false negative* (i.e., that oil does not exist when it does). This situation, which includes actual payoffs from a decision to drill or not drill, is depicted in Fig. 22.6.

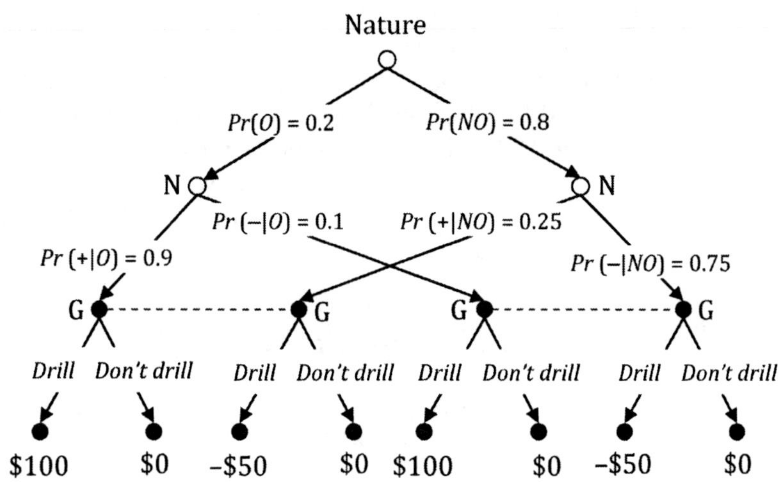

Fig. 22.6.

If GLOMAR drills and strikes oil, it will earn $100 million. If GLOMAR drills and does not strike oil, it will lose $50 million. The expected payoff from drilling without seismic testing is 0.20($100) + 0.80(–$50) = –$20 million. Since the payoff from not drilling is $0, risk-neutral GLOMAR will not drill.

How will seismic testing affect this decision? To answer this question, we must ask how seismic testing changes the probabilities used to estimate the expected payoff from drilling. Specifically, we need to calculate the conditional probability that oil is present given positive seismic testing, that is $Pr\ (O \mid +)$. To accomplish this, we will apply Bayes' theorem.

Bayes' theorem allows us to update the probability of an event after learning that some other event has occurred. We denote the *unconditional* probability of event B_i as $Pr\ (B_i)$. Since this probability comes first, it is referred to as a *prior probability*. The probability that event B_i occurs will be revised once we learn that event A has occurred. We refer to this revised probability as a *posterior probability*. We will denote this posterior probability as $Pr\ (B_i \mid A)$. Bayesian updating involves replacing prior probabilities with posterior probabilities. To transform prior probabilities into posterior probabilities we need **Bayes' theorem**, which is summarized in Equation 22.1.

$$Pr(B_i|A) = \frac{Pr(A|B_i) \cdot Pr(B_i)}{\sum_{i=1}^{k} Pr(A|B_i) \cdot Pr(B_i)}. \tag{22.1}$$

The assignment of conditional probabilities is referred to as a player's **belief profile.** A **perfect Bayesian equilibrium** must be consistent with these beliefs. A player must be able to calculate the expected payoffs associated with a move from any decision node in an information set. Since these beliefs are likely to change during the course of game play, it must also be possible for players to revise their strategies and expected payoffs using Bayesian updating whenever possible. On the basis of these conditional probabilities, should GLOMAR drill for oil?

Using a Harsanyi transformation, Nature moves first by assigning probabilities to the presence or absence of oil, which are $Pr\ (O) = 0.2$ and $Pr\ (NO) = 0.8$, respectively. Nature then assigns probabilities to the effectiveness of seismic testing. Suppose it is known that the conditional probability of a false negative when oil is present is 10 percent, that is $Pr\ (-|O) = 0.10$, which implies that the conditional probability of a true positive of $Pr\ (+|O) = 0.90$. Additionally, suppose the conditional probability of a false positive is 25 percent, that is $Pr\ (+|NO) = 0.25$, which implies a true negative of $Pr\ (-|NO) = 0.75$. Using these values, the conditional probability that oil is present given a positive seismic result is

$$Pr(O|+) = \frac{Pr(+|O) \cdot Pr(O)}{Pr(+|O) \cdot Pr(O) + Pr(+|NO) \cdot Pr(NO)}$$
$$= \frac{0.9(0.2)}{0.9(0.2) + 0.25(0.8)} = 0.474. \qquad (22.2)$$

Eq. (22.2) says that when seismic testing is positive, there is a 47 percent probability that GLOMAR will find oil and a 53 percent probability that it will not. Using these posterior probabilities, the expected payoff from drilling is 0.474($100) + 0.526(–$50) = $21.1. Since the expected payoff from not drilling is $0, a risk-neutral GLOMAR will *drill*. By using Bayes' theorem and updating the prior probability of finding oil using seismic testing, the expected value of finding oil has increased from –$20 million to + $21.1 million. As before, although expected payoffs are used to make a drill-don't drill decision, these values should not be confused with the actual payoffs, which will be realized only after the wells have been drilled.[2]

Solved Exercise

Fangorn Aerospace (F) is a defense contractor that would like to hire a welder (W). Fangorn knows that there is a 15 percent chance that the candidate is a qualified *(Q) candidate and an 85 percent chance that the candidate is* not qualified *(NQ), that is, Pr (Q) = 0.15 and Pr (NQ) = 0.85, respectively.*

To reduce the risk of hiring an unqualified welder, Fangorn conducts a pre-employment screening, which indicates that conditional probability of a false positive is Pr (+|NQ) = 0.1 and a true negative is Pr (–|NQ) = 0.9. The conditional probability of a false negative is Pr (–|Q) = 0.2 and a true positive is Pr (+|Q) = 0.8.

Fangorn's decision to hire depends on the expected value of the welder's marginal product less the wage rate and the pre-employment screening. This situation with actual payoffs is

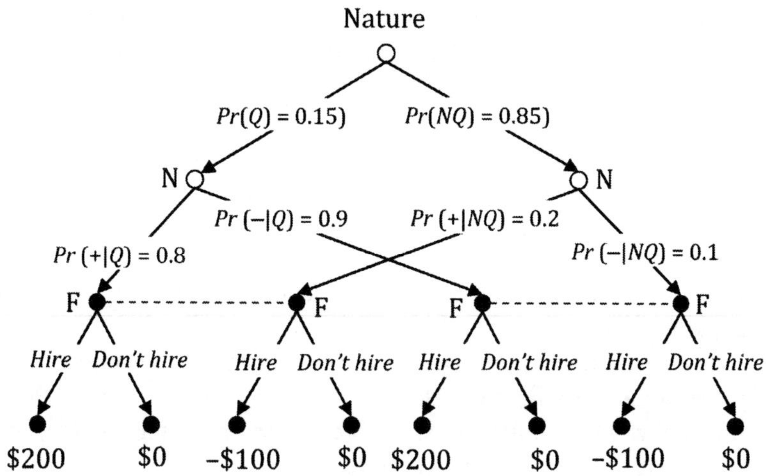

Fig. 22.7.

summarized in the lower branches of Fig. 22.7. If Fangorn hires a qualified candidate it will earn $200. If it hires an unqualified candidate is will lose $100 since pre-employment screening is expensive. Should Fangorn hire this candidate?

Solution

Without pre-employment screening, Fangorn will not hire a welder since the expected payoff is 0.15($200) + 0.85(–$100) = –$22. Using Bayes' theorem, the posterior probability of hiring using pre-employment screening is

$$Pr(Q|+) = \frac{Pr(+|Q) \cdot Pr(Q)}{Pr(+|Q) \cdot Pr(Q) + Pr(+|NQ) \cdot Pr(NQ)}$$
$$= \frac{0.8(0.15)}{0.8(0.15) + 0.2(0.85)} = 0.414. \tag{22.3}$$

From this, Fangorn will hire since the expected payoff is 0.414($200) + 0.586(–$100) = $24.2.

Practice Exercises

22.1 Explain the difference between a valid and an invalid information set.

22.2 Explain how a Harsanyi transformation transforms a static game with incomplete information into a static game with imperfect information. Is it possible to find a Bayesian Nash equilibrium for such static games? Why?

22.3 Explain how a Harsanyi transformation transforms a dynamic game with incomplete information into a static game with imperfect information. Is it possible to find a perfect Bayesian equilibrium for such games? Why?

22.4 Dynamic games with incomplete information that have been transformed into games with imperfect information using a Harsanyi transformation do not have proper subgames. How does this affect our ability to find an equilibrium strategy profile using the backward induction solution algorithm?

22.5 In dynamic games with incomplete information, the pattern of a player's prior moves may reveal private information to his or her opponent, and *vice versa*. Describe the process by which a rational player will revise his or her information set based on this revealed information.

22.6 Dynamic games with incomplete information typically have only one subgame. Explain.

22.7 Explain why implausible Nash equilibria in dynamic games with imperfect information are subgame perfect.

22.8 Explain why it is difficult to find an equilibrium strategy profile for dynamic games with incomplete information.

22.9 Asymmetric information gives rise to the problems of adverse selection and moral hazard. Explain.

22.10 What is the market for lemons? In what way do you think used-car dealers help reduce the problem of adverse selection?

22.11 How does the lemons problem manifest itself in the labor market?

22.12 Do you think that the lemons problem is more or less severe for financial securities traded in the New York Stock Exchange or in the over-the-counter market? Explain.

22.13 Consider the oil-drilling game depicted in Fig. 22.6.

a. Without seismic testing, should GLOMAR drill for oil if the company believes that there is a 30 percent chance of finding oil?

b. Suppose GLOMAR conducts seismic testing and determines that the conditional probability of a positive test $Pr\ (+|O) = 0.75$, and the conditional probability that the test is negative when no oil exists is $Pr\ (-|NO) = 0.40$. Should the company drill for oil?

Notes

1. We earlier defined games involving imperfect information as games in which players move without knowledge of a rival's move. Static games involve imperfect information in the sense that players know their own strategies, but not the strategies of rivals. In this chapter we will expand our definition of imperfect information to include strategic situations involving asymmetric information.

2. Conditional probabilities using Bayes' theorem are very sensitive to changes in prior probabilities. Suppose that the prior probability of finding oil is 10 percent instead of 20 percent. In this case, the posterior probability of finding oil is only 28.6 percent and the expected payoff of finding oil is –$7.1 million. With this small reduction in the probability of oil, risk neutral GLOMAR would choose not to drill. This result is somewhat surprising since seismic testing suggests that there is a strong likelihood that oil is present. It is left as an exercise for the reader to demonstrate that a less accurate test will result in an even larger expected loss from drilling.

CHAPTER TWENTY-THREE

Auctions

Introduction

In bargaining games, buyers and sellers negotiate the terms of a contract. In an **auction**, buyers and sellers compete for the right to purchase or sell something of value. Auctions have been around for thousands of years. The Greek historian Herodotus described auctions involving the sale of brides in Babylonia over 2,500 years ago. A century and a half ago, slave auctions were commonplace in the United States. Today, auctions are used to sell just about everything, including U.S. Treasury bills, equity shares in initial public offerings, tobacco, construction contracts, off-shore drilling rights, airport landing rights, pollution emission rights, service and repair contracts, and a multitude of other goods and services.

The practical significance of game theory in the analysis and design of auctions captured the public's attention in 1994 when the U.S. Federal Communications Commission (FCC) auctioned ten nationwide licenses for "narrow-band" frequencies used by cellular phones, pagers, and other personal communication service (PCS) devices (McMillan, 1994).[1] The auction raised more than $600 million—10 times greater than the most optimistic predictions. The FCC has since used similar auctions to allocate regional broadband licenses for more advanced PCS devices, which sold for more than $8 billion.

Types of Auctions

In this chapter, we will formulate optimal bidding strategies for the most common types of auctions. Optimal bidding strategies depend on the auction's rules and its environment. **Auction rules** refer to the procedures that govern the bidding process, including who may bid, which bids are acceptable, the manner in which bids are submitted, when the auction ends, how the winner is determined, and the price paid. **Auction environment** refers to the characteristics of bidders, including their private value estimates of the object being auctioned, what they know about the private valuations of rival bidders, and when they know it, whether these private valuations are correlated, and the bidders' attitudes toward risk.

There are two basic types of auctions. In a **standard auction**, multiple buyers compete for right to purchase something of value. The bidder with the highest offer price wins. In a **procurement (reverse) auction**, multiple vendors compete for the right to sell something of value. In a procurement auction, the bidder with the lowest asking price wins.

Auctions may be open-bid or closed-bid. In an **open-bid auction**, anyone may participate. In a **closed-bid auction**, participation is by invitation. Bidders may be licensed, required to pay a nonrefundable entry fee, or post a performance bond to ensure timely payment.

Auctions may include a reserve price. In a standard auction a seller's reserve price is the minimum acceptable offer. In an ascending bid-auction, the seller's reserve price may be the opening bid announced by the **auctioneer**, who is the seller's agent. The seller may also have a "secret reserve." If the highest bid is less than the unannounced reservation price, the auction is cancelled. Reservation prices are used to prevent bidder collusion ("bidding rings"), or when low bidder turnout results in a price that is below the minimum value placed on the object by the seller.

Bids may be sealed or oral. In a **sealed-bid auction**, bids are submitted in private. A winner is declared after the bids are "opened.". Sealed bids are typical of many government auctions and are used to sell such things as timber harvesting rights on public lands. In an **oral auction**, bids are announced publicly.

There are two types of sealed-bid auctions. In a **sealed-bid, first-price auction**, the winner submits, and pays, the highest sealed bid. In a **sealed-bid, second-price auctions,** the winner submits the highest bid, but pays the price submitted by the second-highest bidder. A sealed-bid, second-price auction is also called a **Vickery auction** in honor of Nobel laureate William Vickery (1961), who showed that bidders have an incentive to submit offers that reflect their private valuations of the object being auctioned.

Oral auctions differ from sealed-bid auctions in that the bidders' identities are known as offers are submitted, which conveys important information to rival bidders. Suppose, for example, that a bidder in the auction of an oil painting is a wealthy collector. This may cause rival bidders to modify their private valuations and bidding strategies. Since this may result in a higher price, bidders often employ surrogates, or submit bids by telephone or secure Internet connection to conceal their identities. Even when the bidders' identities are unknown, intense bidding activity may cause rivals to adopt more aggressive auction strategies.

There are two main types of oral auctions. In a standard, **English (open-ascending-bid) auction**, the auctioneer opens the bidding, often at some relatively low price. Sometimes the opening bid is the seller's reservation price. As prices are raised, bidders continue to participate until the asking price exceeds their private valuation, which is referred to as the **dropout price**. The process is repeated until the last bidder remaining is declared the winner.

As prices are incrementally raised, bidders who value the object less drop out. The last bidder remaining is declared the winner. In some cases, the seller retains the right to withdraw the offering if the price does not reach a secret reserve price.

In a standard, **Dutch (open-descending-bid) auction**, the auctioneer begins by announcing a high opening price that no one is expected to accept.[2] As the price is incrementally lowered, bidders decide whether to accept or pass. The first bidder to accept is declared the winner.

Information Structures

In the remainder of this chapter we will develop general principles for optimal bidding strategies under different information structures. Bidders' private valuations reflect a variety of factors, including tastes, preferences, needs, abilities, income, wealth, available information, and expertise. The term **information structure** refers to whether bidders' private valuations are common knowledge, and if so, whether a bidder's private valuation is affected by the private valuations of others. In a **complete-information auction** bidders' private valuations are common knowledge. We would expect, however, that auctions are characterized by *asymmetric information* in which the bidders know their own private valuations, but not the private value estimates of others. This is referred to as an **incomplete-information auction**.

In an incomplete-information auction, bidders' private valuations may be independent or correlated. Bidders are said to have **independent private values** if their private value estimates are unaffected by the private value estimates of other bidders. On the other hand, **correlated value estimates** describes an auction environment in which bidders' private valuations affect, and are affected by, the private value estimates of other bidders. For example, suppose that a previously unknown drawing by Leonardo da Vinci is being sold at auction. Bidders initially disagree about the drawing's authenticity. Value estimates are correlated if the bidders revise upwards their value estimates when an acknowledged da Vinci expert submits a high bid.

Complete-Information Auctions

In this section we will develop optimal bidding strategies where the auction environment is characterized by complete information. Although unrealistic, analyzing complete-information auctions provides useful insights into more realistic bidding scenarios. We will also assume that bidders are risk neutral, which means that bidding strategies are based only on expected payoffs. This assumption simplifies our search for an optimal bidding strategy.

While most individuals are risk averse in the sense that they prefer a sure thing to a risky payoff with the same expected value, the assumption of risk neutrality is approximately satisfied when bidders purchase insurance or diversify by participating in multiple auctions. Still, risk-averse behavior in the formulation of optimal bidding strategies cannot be completely ignored. For this reason, we will return to this issue later in the chapter.

Sealed-Bid, First-Price Auction

In general, bidders and sellers try to maximize their respective positive net benefits, which is the difference between their private valuation and the price. Suppose, for example, that Zoe and Jacob compete for the right to purchase an antique pocket watch in a sealed-bid, first-price auction. Jacob knows that the maximum price that Zoe will pay is $562 and Zoe knows that the maximum price that Jacob will pay is $442. Each bidder would like to pay the lowest price that is below private valuations. The winner submits the highest bid. To speed up the analysis, we will assume that all bids are in $50 increments and that the opening bid is $300, which is the seller's reservation price.

Denote Zoe's bid as b_Z and Jacob's bid as b_J. If Zoe wins, her payoff is $\$562 - b_Z$, and zero if she loses. If Jacob wins, he receives a surplus of $\$442 - b_J$, and zero if he loses. Zoe will never submit a bid that is greater than $562 and Jacob will never submit a bid that is greater than $442. In other words, bids greater than the players' reservations prices are strictly-dominated strategies. If Zoe and Jacob submit the same bid (*b*), and the winner is determined by the flip of a fair coin, then Zoe's expected payoff is 0.5($562 – *b*) and Jacob's expected payoff is 0.5($442 – *b*). The normal form of this one-time, static, competition game is depicted in Fig. 23.1. The Nash equilibrium strategy profile for this game is {$400, $450}, with payoffs of $562 – 450 = $112 and $0 for Zoe and Jacob, respectively.

		Zoe					
		$300	*$350*	*$400*	*$450*	*$500*	*$550*
	$300	(71, 131)	(0, **212**)	(0, 162)	(**0**, 112)	(**0**, 62)	(**0**, 12)
Jacob	*$350*	(**92**, 0)	(**46**, 106)	(0, **162**)	(**0**, 112)	(**0**, 62)	(**0**, 12)
	$400	(42, 0)	(42, 0)	(**21**, 81)	(**0**, **112**)	(**0**, 62)	(**0**, 12)

Payoffs: (Jacob, Zoe)

Fig. 23.1.

The $50 difference between the players' bids reflects the fact that Zoe and Jacob will not submit a bid that is greater than their private valuations. If the bidding increment is $10, Jacob will bid $440 and Zoe will bid $450. If the bidding increment is $0.01, the Nash equilibrium strategy profile is {*$442, $442.01*}. As the bidding increments approach zero, the bid submitted by the winner approaches the private valuation of the next-highest bidder.

The optimal bidding strategy in a complete-information, sealed-bid, first-price auction is for the bidder who values the object most to submit a bid that is greater than the private valuation of the next-highest bidder by the allowable bidding increment. All other bidders should submit bids that yield the smallest non-negative surplus. In Fig. 23.1, Zoe should bid $450 > $442 and Jacob should bid $400 < $442. This complete-information, sealed-bid, first-price auction has a unique Nash equilibrium.

Optimal bidding strategy in a complete-information, sealed-bid, first-price auction: The bidder with the highest private valuation wins by submitting a bid just above the private valuation of the next-highest bidder by the minimum allowable bidding increment. All other bidders should submit bids that yield the smallest non-negative surplus.

Sealed-Bid, Second-Price Auction

In a sealed-bid, second-price auction, the winner submits the highest bid, but pays the next-highest bid. Suppose, for example, that Zoe submits a bid of $500 and Jacob submits a bid of $400. Zoe wins the auction, but pays Jacob's bid of $400. Zoe's payoff is $562 – $400 = $162, compared with $562 – $450 = $112 in a sealed-bid, first-price auction. The payoffs for this game are summarized in Fig. 23.2.

		Zoe					
		$300	*$350*	*$400*	*$450*	*$500*	*$550*
	$300	(71, 131)	(0, **262**)	(0, **262**)	**(0, 262)**	**(0, 262)**	**(0, 262)**
Jacob	*$350*	**(142,** 0)	(46, 106)	(0, **212**)	**(0, 212)**	**(0, 212)**	**(0, 212)**
	$400	**(142,** 0)	**(92,** 0)	(21, 81)	**(0, 162)**	**(0, 162)**	**(0, 162)**

Payoffs: (Jacob, Zoe)

Fig. 23.2.

An examination of Fig. 23.2 reveals an interesting feature of sealed-bid, second-price auctions. Unlike the game in Fig. 23.1 in which Zoe's optimal bidding strategy was $450, in this case her dominant strategy is any bid of $450 or greater. This is because Zoe does not pay her bid when she wins. By contrast, Jacob's strictly-dominant strategy is to submit a bid of $400 since any higher bid results in a negative payoff. As the bidding increments approach zero, Jacob's optimal bidding strategy converges to his private valuation. The reader should verify that this game has multiple Nash equilibria.

In general, the optimal bidding strategy in a complete-information, sealed-bid, second-price auction is for the bidder who values the object most to submit any bid that is greater than the private valuation of the next-highest bidder. All other bidders should submit bids that minimize their non-negative payoffs. If there are no bidding increments, all other bidders should bid their private valuations.

Optimal bidding strategy in a complete-information, sealed-bid, second-price auction: The bidder with the highest private valuation wins by submitting a bid that is greater than the private valuation of the next-highest bidder. All other bidders submit bids that minimize their non-negative payoff. If there are no bidding increments, the winner pays the private valuation of the next-highest bidder.

English Auction

In an English auction, passing as a price is raised does not preclude the bidder from accepting an even higher price. To simplify the analysis, however, we will examine the Japanese version of an English auction in which passing on a price precludes a bidder from reentering at a higher price. Consider the complete-information English auction of the antique pocket watch depicted in Fig. 23.3. As before, Zoe's reservation price is $562; Jacob's reservation price is $442; and there are $50 bidding increments. To simplify the analysis, the seller's reservation price is $400. This game has two stages. Each stage is a static game in which the bidders accept or pass on the current asking price.

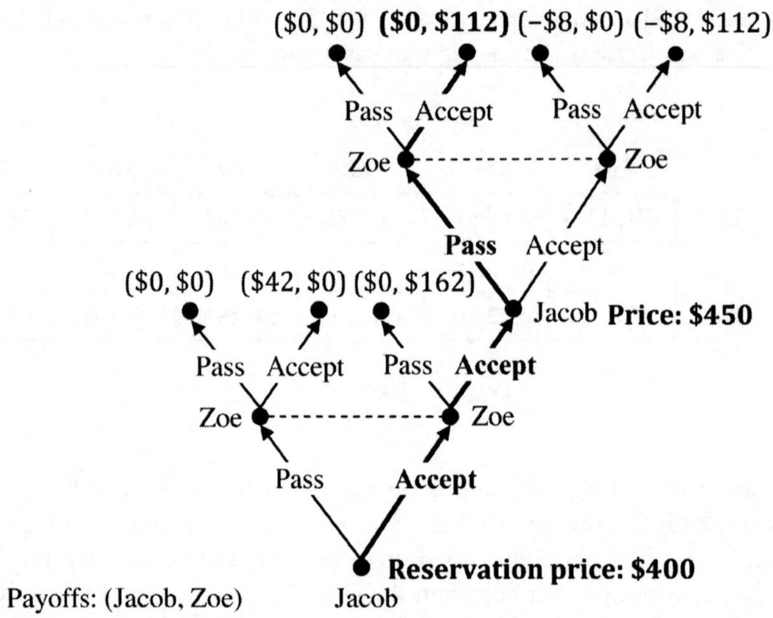

Fig. 23.3.

The auction begins with an opening bid of $400. The static nature of this stage is indicated by the information set that incorporates Zoe's decision nodes. At the opening bid, Zoe and Jacob have a strictly-dominant *accept* strategy since they enjoy positive payoffs of $162 and $42, respectively. In the second stage, the auctioneer raises the price to $450. Jacob's dominant strategy is *pass* since accepting results in a payoff of $442 – $450 = –$8. Zoe's dominant strategy is *accept* since this results in a payoff of $562 – $450 = $112. Since Jacob passes and Zoe accepts, the game ends and the payoffs are $0 and $112, respectively.

In this game, Jacob will continue to participate until the price reaches $400. Jacob's dropout price is $450, which Zoe will accept. If the bidding increment is $0.01,

Jacob will drop out at $442.01, which Zoe will accept. As the bidding increments approach zero, Zoe's optimal bidding strategy approaches Jacob's private valuation of $442.

Optimal bidding strategy in complete-information English auction: All bidders accept prices with non-negative payoffs. The bidder with the highest private valuation wins by accepting the dropout price of the bidder with the next-highest private valuation. If there are no bidding increments, the winner pays the private valuation of the next-highest bidder. A complete-information English auction is strategically equivalent to a complete-information, sealed-bid, second-price auction.

Dutch Auction

To simplify our analysis of a complete-information Dutch auction of the antique pocket watch, we will assume an opening bid of $500. This bid is greater than Jacob's private value estimate, but lower than Zoe's. This game is depicted in Fig. 23.4.

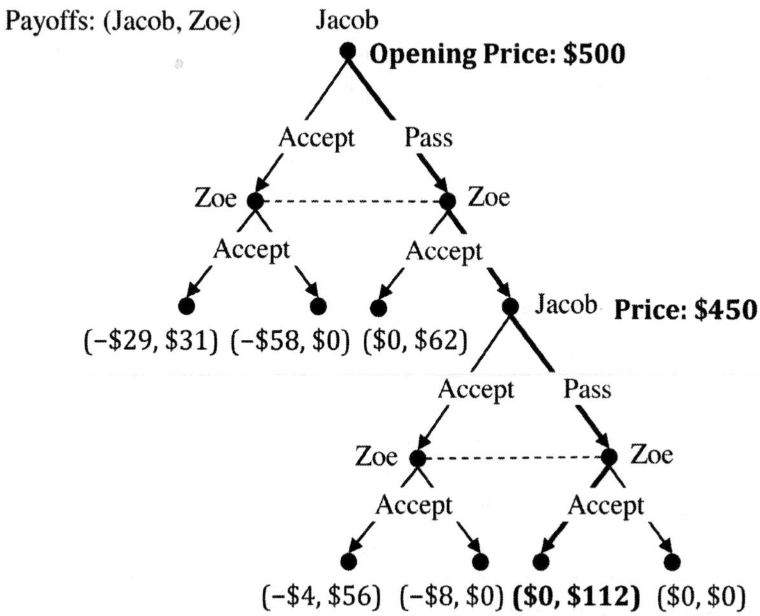

Fig. 23.4.

Zoe knows that Jacob will pass at the opening bid since this price exceeds his private value estimate. Zoe also knows that Jacob will pass when the price is lowered to $450 for the same reason. Jacob will accept, however, when the price is lowered to $400 since this will result in a payoff of $42. Even though Zoe can earn a positive surplus of $62 if she accepts the opening bid, she will wait until the price is lowered to $450 to realize a surplus of $112. If the bidding increment is $0.01, Zoe would wait until the price is lowered to $442.01 and earn $119.99. If there is no bidding increment, Zoe wins and pays Jacob's private value estimate.

Optimal bidding strategy in complete-information Dutch auction: The bidder with the highest private valuation wins by accepting a price that is above the private valuation of the next-highest bidder by the minimum allowable bidding increment. If there are no bidding increments, the winner pays the private valuation of the bidder with the next-highest valuation. This is strategically equivalent to the optimal bidding strategy in a complete-information, sealed-bid, first-price auction.

Expected Revenues from Complete-Information Auctions

In which of these complete-information auctions does the seller receive the highest price? This is an important consideration since the seller determines the auction format. In our examples, we assumed \$50 bidding increments. The prices paid by Zoe were \$450 (sealed-bid, first price), \$450 (English), \$450 (Dutch), and \$440 (sealed-bid, second price). If the size of the bidding increment is \$0.01, the prices paid by the winner are \$442.01(sealed-bid, first price), \$442.01 (English), \$442.01 (Dutch), and \$442 (sealed-bid, second price). If there are no bidding increments, the winner in complete-information, sealed-bid, first- and second-price, English and Dutch auctions is the bidder with the highest valuation and pays what the object is worth to the next-highest bidder, which in our example is \$442.

Revenue-equivalence for complete-information auctions with independent private values: The winner in sealed-bid first- and second-price, English and Dutch auctions has the highest valuation and pays what the object is worth to the next-highest bidder.

Incomplete-Information Auctions with Independent Private Values

Each of the four auction formats examined above assumed an auction environment in which private valuations are common knowledge. This happy state of affairs, however, is rarely encountered in the real world. In this section, we will assume an auction environment characterized by incomplete information. That is, bidders' private value estimates are known only to themselves. Even if this information is inferred, it would not cause bidders to change their private valuations, although it might be used to adjust their bidding strategies.

Sealed-Bid, First-Price Auction

Recall that in a complete-information, sealed-bid, first-price auction the winner is the bidder who values the object most, and submits a bid that is higher by the bidding increment than the private valuation of the next-highest bidder. When information is incomplete, however, each player has an incentive to submit a bid that is somewhat less than what he or she believes the object is worth. While a higher bid increases the likelihood of winning, it also lowers the expected payoff, and vice versa. So what is the bidder's optimal strategy when private valuations are not common knowledge? The answer depends on the number and distribution of the bidders' private valuations.

Although private valuations are not common knowledge, we will assume that private valuations are independent, random, and uniformly distributed, where L is the lowest possible private valuation and H is the highest possible valuation. It has been demonstrated (McAfee and McMillan, 1987) that this auction has a unique, pure-strategy Nash equilibrium in which the optimal strategy of a risk-neutral bidder is

$$b_i^*(V_i) = V_i - \frac{1}{n}(V_i - L), \quad (23.1)$$

where b_i^* is bidder i's optimal bid, V_i is bidder i's private valuation of the object, and n is the number of bidders.

Eq. (23.1) says that a risk-neutral bidder in an incomplete-information, sealed-bid, first-price auction should make an offer that is less than his or her private valuation. How much less depends on the number of bidders. As the number of bidders increases, optimal bids approach a bidder's private valuation. Moreover, as the distribution of private valuations becomes denser, the highest and next-highest valuations converge.

Optimal bidding strategy in an incomplete-information, sealed-bid, first-price auction with independent private values: A bidder should submit a bid that is less than his or her private valuation. This bid approaches the bidder's private valuation with an increase in the number of bidders.

As an illustration, suppose that private valuations are independent, random, and uniformly distributed between L = \$200 and H = \$800. This information may be arrived at in several ways, such by canvassing antique pocket watch experts, surveying Internet sites to identify high and low selling prices, and so on. Applying Eq. (23.1), Zoe's optimal bid is

$$b_Z^*(V_Z) = \$562 - \frac{1}{2}(\$562 - \$200) = \$381. \quad (23.2)$$

Jacob's optimal bid is

$$b_J^*(V_J) = \$442 - \frac{1}{2}(\$442 - \$200) = \$321. \quad (23.3)$$

Zoe and Jacob submit bids that are strictly less than their private value estimates. Suppose that a third bidder, Luke, enters the auction. Luke's private valuation of the pocket watch is V_L = \$386. According to Eq. (23.1), the bidders' optimal bids are b_Z^* = \$441.33, b_J^* = \$361.33 and b_L^* = \$321.00. If there are 20 bidders, these optimal bids become $b_Z^* = \$543.90$, $b_J^* = \$429.90$ and $b_L^* = \$376.70$, and so on. As the number of bidders increases, the optimal bids approach the bidders' private valuations.

Sealed-Bid, Second-Price Auction

In a sealed-bid, second-price auction, the winner submits the highest bid, but pays the bid submitted by the next-highest bidder. In a *complete-information auction* environment, the bidder with the highest private valuation should submit *any* bid that is greater than the private valuation of the next-highest bidder. How does this bidding strategy change in an *incomplete-information auction* environment with independent private values? Since the bidder with the highest private valuation is unknown, each bidder should offer his or her private valuation. There is no incentive for any bidder (including the unknown bidder with the highest valuation) to submit a higher bid.

> *Optimal bidding strategy in an incomplete-information, sealed-bid, second-price auction with independent private values: A bidder should submit a bid that is equal to his or her private valuation. All other bids are strictly dominated.*

It is easy to see why this is an optimal bidding strategy. Suppose that Zoe submits a bid that is greater than her valuation and Jacob submits a bid that is equal to his valuation. Zoe wins the auction and pays Jacob's bid. In this case, Zoe earns a positive surplus. On the other hand, suppose that Zoe outbids Jacob, who submits a bid greater than his private value estimate. Zoe wins the auction and pays Jacob's bid, which may be greater than her private valuation, in which case she suffers a loss. Submitting a bid that is lower than a bidder's private valuation reduces the chance of winning without increasing the payoff. Submitting a higher bid increases the chance of winning, on the other hand, but could result in a negative payoff.

English Auction

What effect does the assumption of independent private valuations have on optimal bidding strategies in an English auction? No effect at all. Recall that in an English auction, prices are announced in ascending order. Bidders should continue to bid until the announced price exceeds his or her dropout price. As bidders drop out, they reveal their approximate private value estimates. Because these estimates are independent, this does not provide information about the dropout prices of the remaining bidders, although the price paid by the winner converges to his or her private valuation. Thus, the optimal strategy in an incomplete-information English auction with independent private values is strategically equivalent to a sealed-bid, second-price auction under the same condition.

> *Optimal bidding strategy in an incomplete-information English auction with independent private values: An incomplete-information English auction with independent private values is strategically equivalent to a complete-information, sealed-bid, second-price auction.*

Dutch Auction

We saw that the optimal bidding strategy in a complete-information Dutch auction is strategically equivalent to a sealed-bid, first-price auction. The winner with the highest valuation accepts a price that is marginally higher than the valuation of the next-highest bidder. In an incomplete-information auction with independent private valuations, a bidder's optimal strategy is to accept a price that is less than his or her private valuation. If these private valuations are independent, random, and uniformly distributed, the optimal bidding strategy is given by Eq. (23.1).

> *Optimal bidding strategy in an incomplete-information Dutch auction with independent private values: An incomplete-information Dutch auction with independent private values is strategically equivalent to an incomplete-information, sealed-bid, first-price auction with independent private values.*

A variation of a standard Dutch auction is the *multiple-item Dutch auction* in which bidders submit sealed offers for blocks of an identical item. The seller accepts the highest price that will dispose of the entire offering, with all bidders paying that price. Multiple-item Dutch auctions are frequently used to sell financial securities, such as U.S. Treasury bills and equity shares in initial public offerings.

> *Multiple-item Dutch auction bidders submit bids for blocks of an identical item. The seller accepts the lowest price that will dispose of the entire offering.*

The optimal bidding strategy in a multiple-item Dutch auction is strategically equivalent to a standard Dutch auction. To illustrate, suppose that the XYZ company announces that it will accept block bids on 1 million shares using a multiple item Dutch auction. The highest bids and the number of shares desired are summarized in Table 23.1.

Table 23.1.

Investor	*Bid per Share*	*Shares*	*Shares Remaining*
A	$10,000	1,000	999,000
B	$100	250,000	749,000
C	$50	150,000	599,000
D	$30	500,000	99,000
E	$25	100,000	−1,000
F	$24.99	50,000	0

The lowest bidder for whom shares are available is investor E who makes an offer to purchase 100,000 shares at $25 per share. Every investor who offers more than $25 receives the number of shares requested, but pays the bid submitted by investor E. Once these requests are satisfied, however, there are only 99,000 shares remaining, which will be allocated to investor E. While bidding high will guarantee that an offer will be accepted, bidding too low runs the risk of not receiving the desired number of shares, or not receiving any shares at all. This was the fate with investor F who bid just 1 cent below the bid submitted by investor E.

Even though everyone pays the lowest bid that disposes the entire offering, submitting a high price does not constitute an optimal strategy. If it did, even the lowest bidder would submit a very high price, which could result in substantial losses for everyone once the shares are sold in the secondary market. This is one reason why sellers frequently specify a price range for acceptable bids. This price range should reflect the sentiments of the market. If the private valuations of a large number of bidders are independent, random, and uniformly distributed, the lowest bid should approximate the lower end of this price range.[3]

Expected Revenues from Incomplete-Information Auctions with Independent Private Values

With incomplete-information auctions and independent private values, bidders do not learn anything about the private valuations of rival bidders. In each of the auctions examined, the winner has the highest private valuation. In the antique pocket watch example, Zoe pays $381 in a sealed-bid, first-price auction, $442 in a sealed-bid, second-price and English auctions, and $562 in a Dutch auction. As the number of bidders increase, the private valuation of the next-highest bidder approaches the private valuation of the highest bidder. In the limit, Zoe pays her private valuation of $562 in each of these auctions.

Revenue-equivalence for incomplete-information auctions with independent private values: The expected price paid in incomplete-information, first- and second-sealed-bid, second-price, English and Dutch auctions, with independent private values is the winner's private value estimate.

Incomplete-Information Auctions with Correlated Value Estimates

In an auction characterized by correlated value estimates, a bidder's private valuation affects, and is affected by, the private valuations of the other bidders. For example, Zoe may value the antique pocket watch more if she believes that Jacob values it highly. In other words, Zoe derives value from owning something that others want.

What kind of information would cause Zoe to raise or lower her private valuation? Zoe might acquire information from the bidding process itself. Zoe might interpret a rival's high bid as reflecting superior information about the quality of the pocket watch. In an English auction, intense bidding activity may be a signal that the pocket watch is worth more than Zoe initially believed. Alternatively, lackluster bidding activity might lead Zoe to downgrade her private value estimate.

How are bidding strategies affected when bids are correlated, or when bidders are uncertain about the true value of the object being auctioned? The answer depends upon the type of auction, and whether the bidders can act on new information. Information received once the auction ends has little or no value.

Common-Value Auctions and the Winner's Curse

A special case of correlated value estimates is a **common-value auction** in which bidders have different private estimates of the true, but unknown, value of the object being auctioned. Suppose, for example, that the U.S. Department of the Interior announces its intention to sell oil-drilling rights in Yosemite National Park using a sealed-bid, first-price auction. The problem is that the actual value of these oil reserves is unknown until the oil is extracted. This leads to a phenomenon called the **winner's curse**, which occurs when the bidder with the most optimistic private value estimate wins, but pays a price that is greater than the true value of the object.

To illustrate the winner's curse, suppose that twenty oil companies submit sealed, first-price bids for the right to drill on public land. Before bidding, each company conducts independent seismic and geological tests to assess the size and value of the oil reserve. Suppose that these private value estimates are independent, random, and uniformly distributed. In a sealed-bid, first-price auction, the company with the most optimistic estimate submits the highest bid and wins the oil-drilling rights. The problem is that the other nineteen companies believe that the oil is worth less than the winning bid. This means that the winner probably paid too much.

How might the oil companies mitigate the winner's curse? Suppose, for example, that a company's senior geologist estimates that the crude oil reserve is worth $100 million. The lowest possible value is $0 in the event of a "dry hole" (water instead of oil) and a maximum possible value of $150 million for "light sweet crude" (oil with low-sulfur content). Rather than submit a bid equal to this estimate, the company uses Eq. (23.1) to submit a bid of $100 – 0.05($100 – $0) = $95 million to mitigate the winner's curse. The optimal bidding strategy in a common-value auction is to submit a bid that is less than its private value estimate.

Optimal bidding strategy in a common-value auction: A *bidder should submit a bid that is less than his or her private value estimate.*

The winner's curse is most pronounced in sealed-bid, first-price auctions because it is not possible to learn anything about the other bidders' private valuations until after the bids have been submitted. By contrast, players may be able to infer something about their rivals' private value estimates during the course of a multistage auction. In an English auction, for example, the player's optimal strategy is to participate until the price exceeds the bidder's private value estimate, which may be raised or lowered based on information gleaned during the bidding process.

Depending on the auction rules, the winner's curse may even prevent a transaction from occurring. Consider the following variation of the winner's curse discussed by Max Bazerman and William Samuelson (1983). Suppose that the owner of a company plans to make a take-it-or-leave-it offer for a privately-held company that produces a related product. Both the buyer and seller believe that the merger will increase the target company's value by 50 percent. The problem is *asymmetric information*. The value of the company is known only to the seller.

To make this example more concrete, suppose that the buyer believes that the company is worth between $20 and $100 million. The seller knows the company is worth $50 million. The expected value of the company to the buyer is 1.5[($20 + $100)/2] = $90 million. What is the buyer's best one-time offer?

The seller will accept any offer that is greater than or equal to $50 million. The seller, for example, will accept an offer of $90 million. At that price, however, the buyer reasons that the company is worth $90/1.5 = $60 million. At that price, however, the buyer's expected loss is $30 million. At a revised price of $60 million, the company is worth $60/1.5 = $40 million for an expected loss of $20 million. The only offer that the buyer can make to avoid the winner's curse is $20 million, which the seller will reject. The example underscores the importance of a properly structured auction that takes into consideration player incentives and bidders who act on revised strategies as new information becomes available.

Incomplete-Information Auctions and Risk Aversion

We have thus far assumed that the bidders are risk neutral. In what ways are bidding strategies affected if bidders are risk averse? It turns out that a bidder's attitude toward risk will affect some bidding strategies, but not others. In the case of an English auction, attitudes toward risk play no role. A bidder will continue to participate until the announced price exceeds his or her private value of the object. There is also no change in bidding strategies with a sealed-bid, second-price auction since a bidder will submit a bid that is equal to his or her private valuation.

By contrast, in a sealed-bid, first-price with independent private values, a risk-averse player will submit a bid that is lower than the bid submitted by a risk-neutral bidder according to Eq. (23.1). How much lower depends on the degree of risk aversion. While this lower bid increases the possibility of losing the auction, it also increases the expected payoff. What about auctions involving correlated value estimates? Recall that in these auctions the information revealed during an English auction reduces the winner's curse. Because of this, risk-averse bidders will submit bids that are somewhat higher than in a second-price auction.

Practice Exercises

23.1 In each of the following auctions with complete information and no bidding increments, who wins and what price is paid?

a. Sealed-bid, first-price auction
b. Sealed-bid, second-price auction
c. Standard English auction
d. Standard Dutch auction

23.2 In each of the following auctions with incomplete information, independent private values, and no bidding increments, who wins and what price is paid?
a. Sealed-bid, first-price auction
b. Sealed-bid, second-price auction
c. Standard English auction
d. Standard Dutch auction

23.3 Explain why a player in a sealed-bid, second-price auction would never submit a bid that exceeds his or her private value of the object being auctioned.

23.4 In a Dutch auction with complete information, the winner values the object the most and pays his or her private value of the object. Do you agree? Explain.

23.5 What is the difference between optimal bidding strategies in incomplete-information auctions with independent and correlated private value estimates?

23.6 In a sealed-bid, first-price auction with independent private values, risk-neutral bidders will submit bids that are higher than their private values of the object. Do you agree? Explain.

23.7 In an incomplete-information, sealed-bid, first-price auction with independent private values, each bid will be lowered as the number of risk-neutral bidders increases, eventually converging to the next-highest bidder's valuation of the object being auctioned. Do you agree? Explain.

23.8 In an incomplete-information, sealed-bid, second-price auction with independent private values, risk-neutral bidders should submit bids that are equal to what each believes the object is worth to the next-highest bidder. Do you agree? Explain.

23.9 A Dutch auction with independent private values and risk-neutral bidders is strategically equivalent to a sealed-bid, second-price auction. Do you agree? Explain.

23.10 What is a common-value auction? Explain what is meant by the "winner's curse."

23.11 In what way do bidding strategies differ when winners are risk averse instead of risk neutral?

23.12 Why would a company going public using a multiple-item Dutch auction begin by specifying a price range for investor bids?

23.13 At the Green Dragon tavern Jethro announces that he will auction off an envelope containing $35.00. Clem and Heathcliff are the only two bidders and each has $40 with which to bid. The rules of the auction are as follows:
(1) The bidders take turns. After a bid is made, the next bidder can make either another bid, or pass. The opening bid must be $10.00.
(2) After the opening bid, each bid must be in increments of $10.
(3) Bidders cannot bid against themselves.

(4) The bidding comes to an end when either bidder passes, except on the first bid. If the first bidder passes, the second bidder is given the option of accepting the bid.
(5) The highest bidder wins.
(6) All bidders must pay Jethro the amount of their last bid.
(7) Assume that Clem bids first.
a. Illustrate this auction with a game tree.
b. Determine the subgame perfect equilibrium for this game using the backward induction solution concept.
c. What is the outcome of this auction?

23.14 Suppose that Ted and Bill plan to submit sealed bids for a pair of tickets for a sold-out basketball game between the Orange of Syracuse University and the Wildcats of the University of Arizona. Earlier in the day, Ted overheard Bill tell his girlfriend that he is willing to pay as much as \$212. The previous night, Bill, who is dating Ted's sister, learned that Ted would be willing to pay as much as \$224. In this auction, there is a reservation price of \$200 and bids must be submitted in \$5 increments. Although the winner of this auction submits the highest bid, he will pay the bid submitted by the loser. If Ted and Bill submit the same bid, the winner will be determined by the flip of a fair coin.
a. Illustrate this auction as a noncooperative, simultaneous-move, one-time game.
b. Does Ted or Bill have a dominant strategy?
c. What is the solution profile for this game? Who will win the auction, and how much will he pay?
d. Suppose that the bidding increments were \$0.01. Who will win the auction and how much will he pay?

23.15 The U.S. Department of the Interior has announced its intention to use a sealed-bid, first-price auction to sell petroleum-drilling rights on public land in Alaska. Four oil companies have indicated their intention to participate in this auction. The problem is that the actual value of the oil reserves is unknown with certainty. Before submitting their bids, each oil company conducted independent seismic and geological tests to obtain an estimate of the quality and quantity of the petroleum reserves. The petroleum-deposit value estimates in millions of dollars of the four companies are V_1 = \$250, V_2 = \$170, V_3 = \$320, and V_4 = \$265. Although the companies do not know their rivals' estimates, they believe that these valuations are independent and randomly distributed between a low value of L = \$100 and a high possible price of H = \$400.
a. What is the optimal bid of each oil company?
b. Who wins the auction and what price will the winner pay for the drilling rights?

c. Do you believe that the winner of this auction paid too much for the drilling rights? Why?

23.16 Consider an incomplete-information, sealed-bid, first-price auction in which the players have independent private values. The general consensus is that the true value of the object being sold is uniformly distributed between $1 and $10.

a. If the true value of a risk-neutral player is $2, what is this player's optimal bidding strategy if there are only two bidders?

b. If this is a Dutch auction, what is the player's optimal bidding strategy if there are three bidders?

c. If this is a sealed-bid, second-price auction, what is a player's optimal bidding strategy?

23.17 Consider an incomplete-information, independent private value auction with twenty risk-neutral players. The players' valuations are generally believed to be uniformly distributed between $0 and $50,000. Suppose that the true private value of one player is $40,000. For each of the following auction formats, what is the player's optimal bidding strategy?

a. Sealed-bid, first-price

b. Sealed-bid, second-price

c. English

d. Dutch

Notes

1. After consulting game theorists and auction experts, the FCC devised a multiple-round auction in which four telecommunications companies paid $3.5 million for the right to submit bids for up to three of ten licenses.

2. Dutch auctions, which date back to the nineteenth century, were named after the system of selling tulip bulbs in Holland.

3. On August 19, 2004, Google went public by selling 19.6 million shares at $85 per share using a multiple-item Dutch auction. This initial public offering (IPO) raised $1.67 billion. Google originally specified a price range of $108-$135, which was lowered to $85-$95 because of a weak market. This price range, however, was still too high to guarantee that the IPO would be fully subscribed at 25.6 million shares. Google may have miscalculated, or it may have thought it could earn even more by selling the remaining 6 million shares in the secondary market. In fact, when trading began in the secondary market, Google opened at almost $136 per share. By mid-2013, Google shares were selling for more than $900.

CHAPTER TWENTY-FOUR

Signaling

Introduction

Among the many applications of game theory is an analysis of a process known as **signaling**. Signaling is one possible solution to the adverse selection problem. Signaling occurs when a player transmits a credible and unreproducible signal to convey information to a less informed player. A signal is intended to convey important information that is otherwise not directly observable.

Signals are a common feature of the market place. Companies signal buyers about the benefits of their product, which are otherwise unobservable until the product is consumed. The purpose of these signals is to persuade potential buyers to give their product a try. These signals may include money-back guarantees, free trial periods, and so on. In some instances, signals are linked to a firm's reputation. This "you-know-me-so-you-can-trust-me" approach often entails packaging that prominently displays the company's logo, or can be as simple as a neighborhood storefront proudly displaying the year in which the company was founded.

Spence Education Game

Signaling is characteristic of labor markets. Job seekers know whether they are productive, which is not directly observable by prospective employers. If productivity were directly observable, workers would be paid the value of their marginal product. Since productivity is not observable, workers are offered compensation equal to average productivity. The challenge facing employers is to separate low-productivity and high-productivity workers.

Suppose, for example, that a potential employer believes that there are two types of workers: Low-productivity types who contribute nothing to the firm's annual revenues, and high-productivity types who contribute \$100,000. The employer believes that there is a 50-50 chance of hiring a high-productivity type. The expected value of a worker's marginal product is 0.5(\$0) + 0.5(\$100,000) = \$50,000. For this reason, the most that the employer is willing to offer is \$50,000.

Unfortunately, this situation creates a lemons problem. The applicant knows his or her productivity type, but the employer does not. Suppose that a high-productivity type can start her own business and earn \$75,000 and a low-productivity type can start his own business and earn \$40,000. The high-productivity types will reject the firm's offer, which the low-productivity type will accept. Since the market will become crowded with low-productivity types, the firm's optimal strategy is not to employ anyone.

Is there a solution to this problem? Michael Spence (1973) argued that education could be used as a signal of a worker's productivity type.[1] Suppose that a firm (F) is considering hiring a worker (W) who is asking for a wage (w). The applicant knows the value of her marginal productivity (π). This situation is characterized by asymmetric information. The applicant knows whether she is a high-productivity (H) or a low-productivity (L) type. Productivity is private information that the firm cannot directly observe. The applicant's next best alternative is to open her own business and earn $\theta\pi_H$ if she is a high-productivity type and $\theta\pi_L$ if she is a low-productivity type, where $0 < \theta < 1$.

Suppose that a worker uses observable education as a signal for unobservable productivity. The decision to attend college (C) or to stop after secondary school (S) depends on the opportunity cost of attending college relative to what can be earned by working for someone else. The opportunity cost of attending college is a sunk cost that includes tuition, fees, tutors, textbooks, and foregone earnings. This opportunity cost for low-productivity types (c_L) is greater than the opportunity cost for high-productivity types (c_H), where $c_L > c_H$. This is because it takes longer for a low-productivity type to graduate, either because the student enrolls for few credits each semester, or because they repeat failed courses. Of course, workers who do not go to college do not incur this expense. For simplicity, we will assume that education has no effect on a worker's productivity.

The payoff to the firm that hires a high-productivity type is $\pi_H - w$ and the payoff from hiring a low-productivity type is $\pi_L - w$, where $\pi_H < \pi_L$. The payoff to a high-productivity type who attends college is (the present value of) $w - c_H$. The payoff to a low-productivity type who attends college is $w - c_L$. The payoffs to high-productivity and low-productivity college graduates who are not hired are $\theta\pi_H - c_H$ and $\theta\pi_L - c_L$, respectively. The earnings of high-productivity and low-productivity types who do not attend college and are not hired are $\theta\pi_H$ and $\theta\pi_L$, respectively.

This version of the Spence education game is depicted in Fig. 24.1. The game proceeds as follows. Nature moves by determining the probability distribution of productivity types, where $Pr(H)$ is the probability of a high-productivity type and $1 - Pr(H) = Pr(L)$ is the probability of a low-productivity type. Next, a worker (W) with a college or secondary education applies for a job and asks for a wage (w). The firm (F) either makes a job offer (O), or does not (N). If the firm does not make an offer, the payoff to the applicant is the earnings from self-employment. If the firm makes an offer, the worker either accepts (A) or rejects (R). If the worker rejects, the payoff is the same as if no offer had been made. If the worker accepts, the payoff is the wage less the cost of a college (if any). The payoff to the firm is the value of the worker's marginal product less the wage. The information sets indicate that the applicant knows her productivity type, which the firm cannot directly observe.

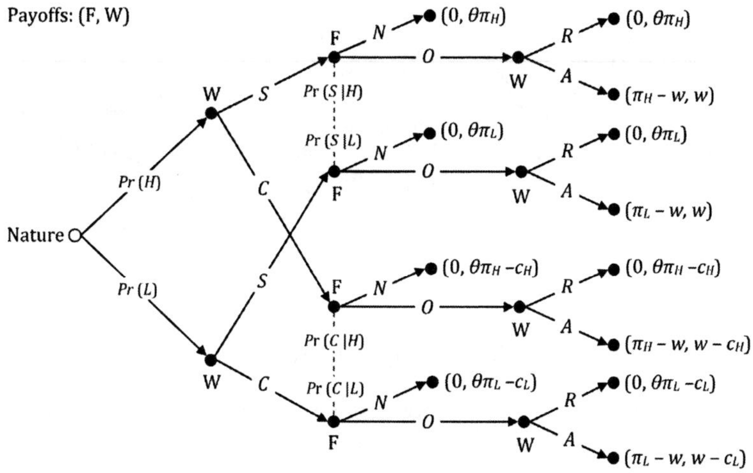

Fig. 24.1.

Pooling Strategy

We begin our analysis by assuming that the firm does not use the education as a productivity signal. The firm lumps all productivity types into the same pool of workers. This **pooling strategy** assumes that $Pr\ (C \mid H) = Pr\ (C \mid L) = Pr\ (H)$ and $Pr\ (S \mid H) = Pr\ (S \mid L) = Pr\ (L)$. Assume that the probability distribution of high-productivity and low-productivity types is $Pr\ (H) = 0.4$ and $Pr\ (L) = 0.6$, respectively. Suppose that the value of the marginal product of high-productivity and low-productivity types are $\pi_H = \$12$ and $\pi_L = \$8$, respectively. The worker's payoff from self-employment is $\theta\pi$, where $\theta = 0.5$. The cost of attending college for high-productivity and low-productivity types are $c_H = \$1$ and $c_L = \$4$, respectively. Suppose the wage rate is $w = \$10$. The payoff to the firm from a high-productivity worker is $\pi_H - w = \$12 - \$10 = \$2$ and the payoff from low-productivity workers is $\pi_L - w = \$8 - \$10 = -\$2$.

Fig. 24.2 summarizes the expected payoffs in this game. For example, the firm's expected payoff from hiring a worker who attended college is $Pr\ (H)(\pi_H - w) + Pr\ (L)(\pi_L - w) = 0.4(2) + 0.6(-2) = -0.4$. While the worker knows her payoff with certainty, the firm's expected payoffs depend on the probability distribution of productivity types. The reader should verify that the subgame-perfect equilibrium strategy profile is for the firm not to make a job offer at the requested wage regardless of education. This outcome is referred to as a **pooling equilibrium**. In this case, the applicant should consider lowering her wage demand.

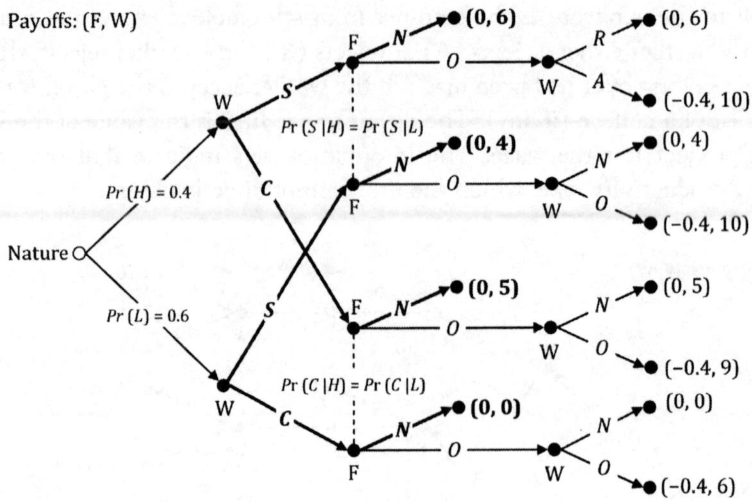

Fig. 24.2.

Separating Strategy

In the above example, the firm does not make a job offer at the requested wage if it adopts a pooling strategy. Is this outcome different when the firm uses education as a signal for productivity type? We will proceed with our search for a subgame-perfect Bayesian equilibrium in two ways. We begin by assuming that the firm uses education as a *perfect signal* for productivity type. That is, the firm assumes that only high-productivity types attend college. Using education in this way to distinguish (separate) high-productivity from low-productivity types is called a **separating strategy.**

Since the firm uses education as a perfect signal of worker productivity, $Pr\ (C\,|\,H) = Pr\ (S\,|\,L) = 1$ and $Pr\ (C\,|\,L) = Pr\ (S\,|\,H) = 0$. That is, the conditional probability that a worker who attended college is a high-productivity type and the conditional probability that a worker who did not attend college is a low-productivity type is unity, and the conditional probability that a worker who attended college is a low-productivity type and the conditional probability that a worker who did not attend college is a high-productivity type is zero.

To find the subgame-perfect Bayesian equilibrium, it is necessary to determine the firm's conditional expected payoffs using Bayes' theorem. The conditional expected payoff to the firm that hires a high-productivity worker who attended college is $Pr\ (H\,|\,C)(\pi_H - w)$, where

$$Pr(H|C) = \frac{Pr(C|H)\cdot Pr(H)}{Pr(C|H)\cdot Pr(H) + Pr(C|L)\cdot Pr(L)} = \frac{1(0.4)}{1(0.4)+0(0.6)} = 1. \quad (24.1)$$

Thus, the conditional expected payoff from hiring a worker who attended college is $Pr\ (H\,|\,C)(\pi_H - w) + Pr\ (L\,|\,C)\ (\pi_L - w) = 1(2) + 0(-2) = 2$. Repeating this procedure for all conditional expected payoffs we obtain Fig. 24.3.

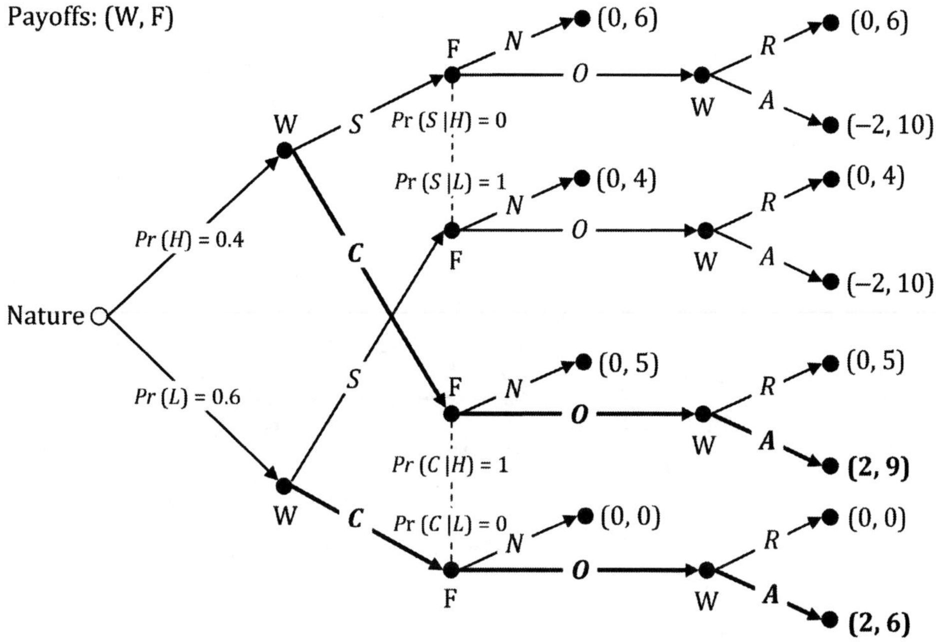

Fig. 24.3.

The reader should verify that the subgame-perfect Bayesian equilibrium from a separating strategy is {C → O → A}. When education is used as a perfect signal for productivity type, the firm will offer employment to college graduates, who will accept since the wage is greater than the payoff from self-employment. This outcome is referred to as a **separating equilibrium**. Of course, the firm's actual payoffs depend on whether it hires a high-productivity or a low-productivity type college graduate.

A separating strategy in which education is used as a perfect signal for a worker's productivity type appears shortsighted since employers realize that some low-productivity types attend college. What if the firm uses education as an *imperfect signal* of worker productivity? Suppose the firm believes that 80 percent of college graduates are high-productivity types and 20 percent are low-productivity types. The firm also believes that 30 percent of workers who did not attend college are high-productivity types and 70 percent are low-productivity types. In this case, education is a strong (albeit not perfect) signal of productivity. Fig. 24.4 summarizes the conditional expected payoffs in which *Pr* (C | *H*) = 0.8, *Pr* (C | *L*) = 0.2, *Pr* (S | *H*) = 0.3, and *Pr* (S | *L*) = 0.7. These conditional beliefs may be summarized as {C: (*H*: 0.8, *L*: 0.2), S: (*H*: 0.3, *L*: 0.7)}.

Perhaps not surprisingly, the subgame-perfect Bayesian equilibrium from a separating strategy from this belief profile is {C → O → A}. This is the same outcome as when education was used as a perfect signal of worker productivity in Fig. 24.3.

Solved Exercise

Suppose that firm A is deciding whether to launch (L) or not launch (NL) a new product. Firm B's decision to launch its own new product depends on whether A launches a weak

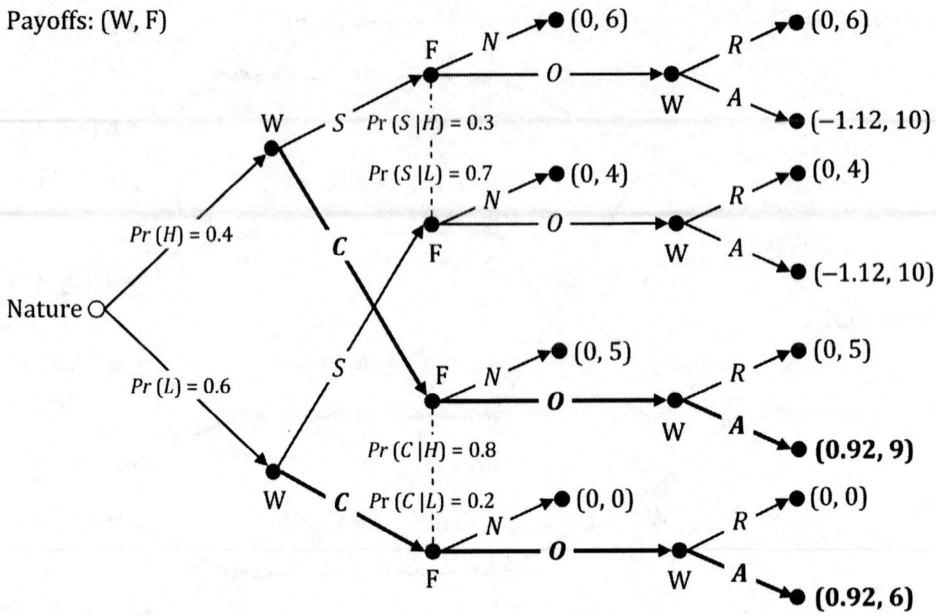

Fig. 24.4.

(W) or a strong (S) product. Firm A knows whether its new product is a weak or strong product, which B can only infer. Firm B believes that there is a 80 percent probability that A has a strong product and a 20 percent probability that A has a weak product. The firms' payoffs are summarized in the following figure.

It is A's best interest if B does not launch. To dissuade B from launching, A "leaks" the results (R) of test data that indicate its product is strong. The problem for firm B is that it does not know whether the report is true, or whether the report has been falsified to make A's weak product appear to be strong. In other words, B does not know if A is "bluffing." Suppose that B believes that there is a 25 percent chance that the report is accurate and a 75 percent chance that the report is a bluff.

a. *What is the Bayesian subgame-perfect equilibrium if firm B adopts a pooling strategy?*
b. *What is the Bayesian subgame-perfect equilibrium if firm B adopts a separating strategy?*

Solution

a. *If B adopts a pooling strategy Pr (R | S) = Pr (R | W), B's expected payoff if A launches a strong product is 0.8(–200) + 0.2(400) = –80. B's expected payoff if A launches a weak product is 0.8(200) + 0.2(400) = 240. The pooling equilibrium depends on whether A launches a weak or strong product. The pooling equilibrium is {L → NL} and {NL → L}. That is, B will not launch if A launches, and will launch if A does not launch.*

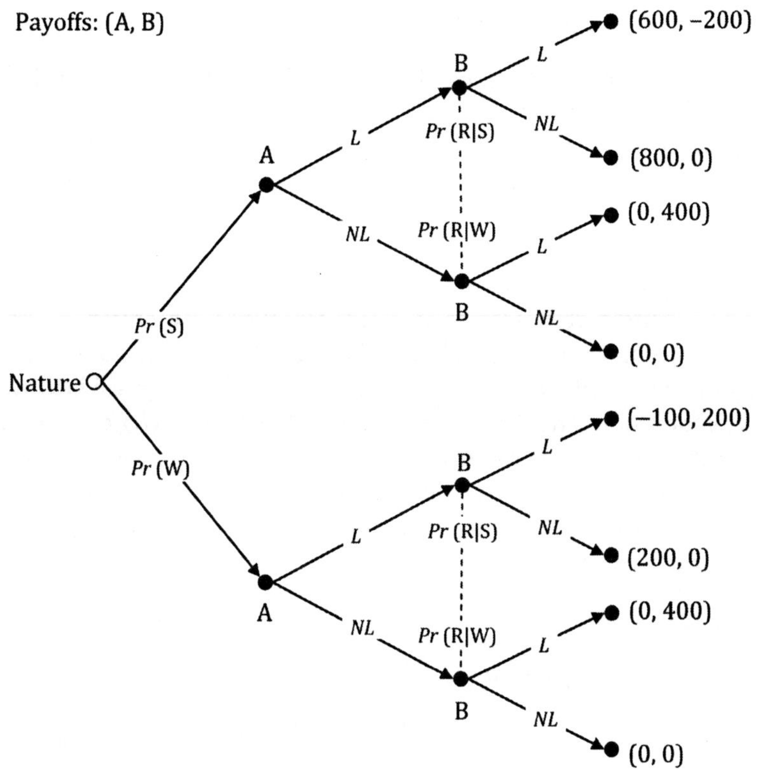

Fig. 24.5.

b. Applying Bayes' theorem,

$$Pr(S|R) = \frac{Pr(R|S) \cdot Pr(S)}{Pr(R|S) \cdot Pr(S) + Pr(R|W) \cdot Pr(W)}$$
$$= \frac{0.25(0.8)}{0.25(0.8) + 0.75(0.2)} = 0.57, \tag{24.2}$$

which implies that Pr (R | W) = 0.43. If B adopts a separating strategy, the expected payoff if A launches a strong product is 0.57(−200) + 0.43(400) = 58. B's expected payoff if A launches a weak product is 0.57(200) + 0.43(400) = 286. The separating equilibrium depends on whether A launches a weak or strong product. The separating equilibrium is {L → L} and {NL → L}. That is, B will launch regardless of whether A launches or does not launch.

Corporate Investment Game

Another example of signaling is the following corporate investment game (Myers and Majluf, 1984; Dybvig and Zender, 1991). In this game, the owner-manager (M) of a company requires investment financing. The manager has private information about the profitability of the investment. Suppose the manager offers a potential shareholder (S) an equity stake ($0 \leq \theta \leq 1$) in the company in exchange for the requisite financing. How much of an equity stake should the manager offer, and under what conditions should the potential shareholder accept?

In this game, the shareholder must decide whether to interpret the size of the offer as a signal of the firm's profitability. Suppose that the company's profit (π) is high (π_H) or low (π_L), where $\pi_H > \pi_L > 0$, which depends on several factors, including market conditions. The decision to accept (A) or reject (R) the offer depends on whether the payoff from the investment (Z) is greater or less than the payoff from the next best alternative investment $I(1 + i)$, where i is the risk-adjusted rate of return. The capital investment game, which is characterized by asymmetric information, is depicted in Fig. 24.6.

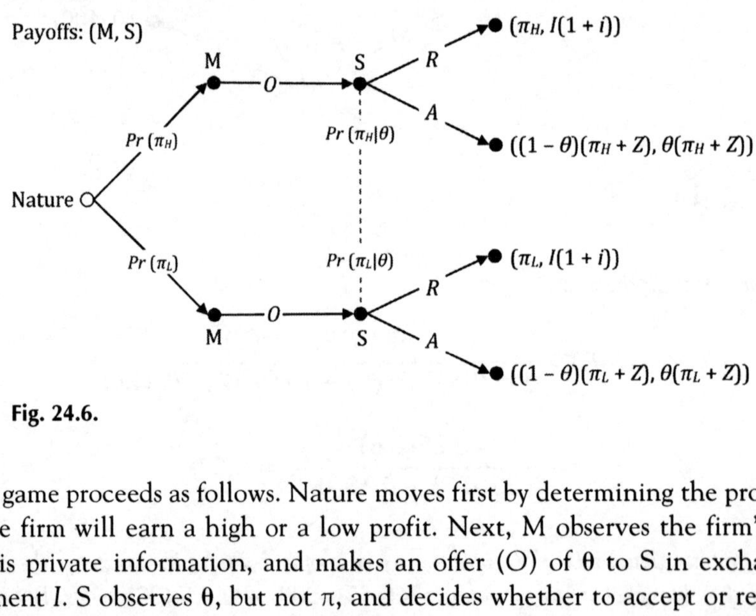

Fig. 24.6.

The game proceeds as follows. Nature moves first by determining the probability that the firm will earn a high or a low profit. Next, M observes the firm's profit, which is private information, and makes an offer (O) of θ to S in exchange for investment I. S observes θ, but not π, and decides whether to accept or reject M's offer. If S rejects M's offer, the payoffs are $(\pi_H, I(1 + i))$ or $(\pi_L, I(1 + i))$ depending on whether the firm earns a high profit or a low profit. If S accepts M's offer, the payoffs are $((1 - \theta)(\pi_H + Z), \theta(\pi_H + Z))$ or $((1 - \theta)(\pi_L + Z), \theta(\pi_L + Z))$. The decision by S to accept or reject M's offer depends on several factors, such as the size of the required investment, the equity share offered, and whether S adopts a pooling strategy or a separating strategy.

If S adopts a pooling strategy then the expected payoff depends only on the probability distribution of profits determined by Nature. Thus, S will accept the offer if

$$\theta[\,Pr(H)\pi_H + Pr(L)\pi_L + Z] \geq I(1 + i). \tag{24.3}$$

The size of M's offer must satisfy the condition

$$\theta[(\pi + Z)] \leq Z. \tag{24.4}$$

Condition (24.3) says that the amount paid to S must not exceed the expected payoff of the investment. Conditions (24.3) and (24.4) may be solved simultaneously to determine the pooling equilibrium for this game. To clarify the issues involved, consider the following concrete example.

Suppose that M offers S an equity share of $\theta = 0.25$ in exchange for an investment of $I = \$4$, which will generate a return of $Z = \$6$. Nature has determined that $Pr(\pi_H = \$15) = 0.3$ and $Pr(\pi_L = \$10) = 0.7$. Finally, S can expect to earn a rate of return on the next best alternative investment of 12 percent ($i = 0.12$). What is the subgame-perfect Bayesian equilibrium if S adopts a pooling strategy? In other words, the investor does not use the size of the manager's offer as a signal for the firm's profit, that is $Pr(\pi_H) = Pr(\theta \mid \pi_H)$. The expected payoffs for this game are summarized in Fig. 24.7.

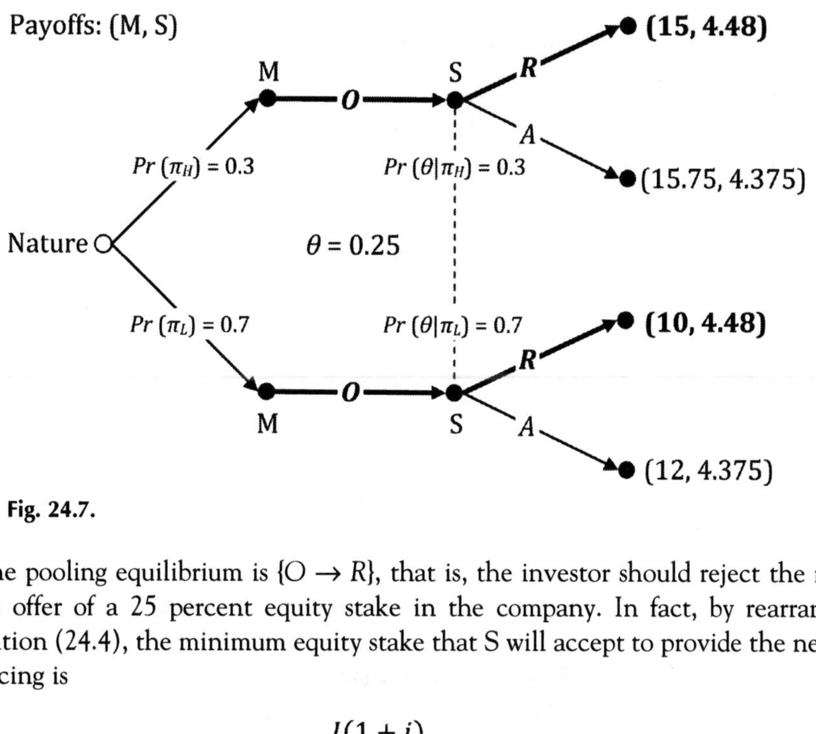

Fig. 24.7.

The pooling equilibrium is {O → R}, that is, the investor should reject the manager's offer of a 25 percent equity stake in the company. In fact, by rearranging condition (24.4), the minimum equity stake that S will accept to provide the needed financing is

$$\theta \geq \frac{I(1+i)}{Pr(\pi_H)\pi_H + Pr(\pi_L)\pi_L + Z} = 0.256, \tag{24.5}$$

or 25.6 percent. Rearranging condition (24.5), the maximum share that the manager will offer is

$$\theta \leq \frac{Z}{(\pi_H + Z)} = 0.286 \tag{24.6}$$

if profits are high and

$$\theta \leq \frac{Z}{(\pi_L + Z)} = 0.375 \tag{24.7}$$

if profits are low.

Although S will reject an offer of 25 percent, conditions (24.6) and (24.7) suggest that M can obtain the financing by sweetening the deal. How might these results differ if S were to use M's offer as a signal for the firm's profitability? In other words, what is the subgame-perfect Bayesian equilibrium if S adopts a separating strategy?

If S adopts a separating strategy, condition (24.3) becomes

$$\theta[Pr(\pi_H|\theta)\pi_H + Pr(\pi_L|\theta)\pi_L + Z] \geq I(1+i), \tag{24.8}$$

where

$$Pr(\pi_H|\theta) = \frac{Pr(\theta|\pi_H) \cdot Pr(\pi_H)}{Pr(\theta|\pi_H) \cdot Pr(\pi_H) + Pr(\theta|\pi_L) \cdot Pr(\pi_L)}. \tag{24.9}$$

Suppose that $Pr\ (\theta\,|\,\pi_H) = 0.9$. In this case, the offer is used by S as a strong, albeit not perfect, signal of the firm's profitability. In this case, $Pr\ (\pi_H\,|\,\theta) = 0.794$ and $Pr\ (\pi_L\,|\,\theta) = 0.206$. The resulting conditional expected payoffs are summarized in Fig. 24.8.

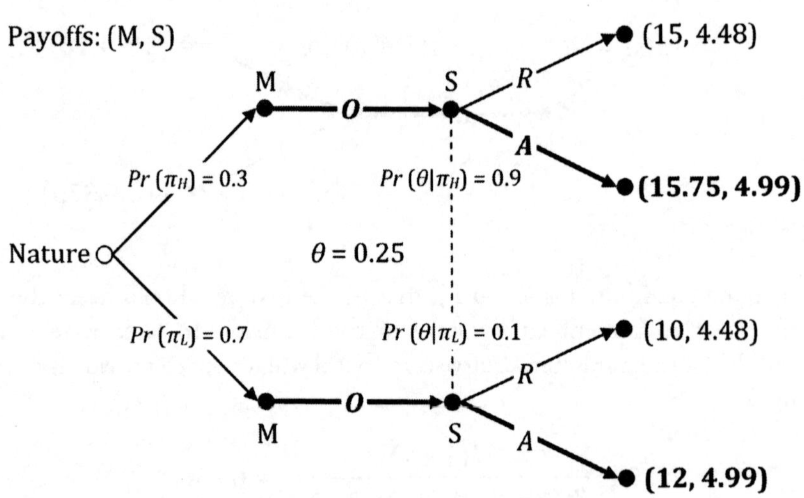

Fig. 24.8.

The separating equilibrium is for the investor to *accept* the manager's offer, that is, {O → A}. In this game, the minimum share that S will accept to provide the needed financing is

$$\theta \geq \frac{I(1+i)}{Pr(\pi_H|\theta)\pi_H + Pr(\pi_L|\theta)\pi_L + Z} = 0.224, \tag{24.10}$$

or a 22.4 percent equity stake, which is less than the manager's 25 percent offer.

Multiple Subgame-Perfect Bayesian Equilibria

The important lesson to emerge from the games examined in this chapter is that signals can be effective in overcoming the problem of asymmetric information, but only if they are differentially costly to send. In the Spence education game, for example, signals were effective because it was less costly for high-quality types to go to college than for low-quality types. Signals were effective in the corporate investment game because the equity share offered was inversely related to the firm's profitability.[2]

Unfortunately, the existence of a subgame-perfect Bayesian equilibrium in each of the signaling games examined depends on the use of a pooling or separating strategy. This is true of almost all signaling games. Even more disturbing is the fact that the choice of strategies may result in conflicting results. In fact, a unique subgame-perfect Bayesian equilibrium is more the exception than the rule. The reason for this is that there are no proper subgames, which are only characteristic of games with complete and perfect information. For the reason, the search for a perfect Bayesian equilibrium can be problematic.

In the signaling games examined in this chapter, backward induction was used to find the subgame-perfect Bayesian equilibrium. In fact, there is no commonly accepted solution algorithm for games involving imperfect information. Whatever method is used, however, the resulting outcome must satisfy three requirements. First, the collection of Nash equilibrium strategy profiles must be consistent with the players' beliefs about the state of nature and the rivals' strategies. Second, players maximize their conditional expected payoffs given the available information. Finally, players should revise conditional expected payoffs using Bayesian updating whenever possible.

> **Principle:** *For a strategy profile and a belief profile to constitute a perfect Bayesian equilibrium, it must satisfy three conditions: (i) Given the players' beliefs, the collection of strategy profiles must constitute a Nash equilibrium; (ii) players attempt to maximize their conditional expected payoffs from each information set, and (iii) conditional expected payoffs will be revised using Bayesian updating whenever possible.*

Although it is easy to define the necessary conditions for a perfect Bayesian equilibrium, it can be difficult to identify them in practice. One possible approach involves three steps. The first step is to propose a set of strategies and beliefs. Next, check whether these proposed strategies satisfy conditions (i) and (ii) in the above principle. Lastly, check to see if these beliefs satisfy requirement (iii). The main problem with this approach is that there are typically a very large number of possible strategies and belief profiles, which can make the search for a perfect Bayesian equilibrium difficult, time consuming, and, quite possibly, impossible.

Practice Exercises

24.1 In terms of signaling, explain why graduates of Ivy League universities receive higher starting salaries than do graduates from less prestigious institutions. Why is this a problem of adverse selection, and do you believe that this outcome is fair? Explain.

24.2 In your opinion, what is a better signal for worker productivity: A so-so grade point average from a prestigious research university, or a high grade point average from a mediocre private or public college or university? Regardless of your answer, what other signals should job applicants provide to improve their chances of employment?

24.3 Consider the Spence education game depicted in Fig. 24.4. Suppose that the firm's belief profile about education is {C: (*H*: 0.80, *L*: 0.20), S: (*H*: 0.50, *L*: 0.50)}.

a. Use Bayes' theory to calculate conditional expected payoffs. Illustrate the extensive form of this game.
b. What is the perfect Bayesian equilibrium if the manager adopts a pooling strategy?
c. What is the perfect Bayesian equilibrium if the manager adopts a separating strategy?
d. Compare your answers to parts b and c. What, if anything, can you say about the existence of a unique perfect Bayesian equilibrium in signaling games? Be sure to mention the importance of the firm's beliefs, states of nature, and the signaling strategy used.

24.4 Consider the Spence education game depicted in Fig. 24.4. Suppose that the firm's belief profile is {C: (*HP*: 0.20, *LP*: 0.80), S: (*HP*: 0.20, *LP*: 0.80)}.

a. Use Bayes' theory to calculate conditional expected payoffs. Illustrate the extensive form of this game.
b. What is the perfect Bayesian equilibrium from a pooling strategy?
c. What is the perfect Bayesian equilibrium from a separating strategy?
d. Compare your answers to parts b and c. What, if anything, can you say about the existence of a unique perfect Bayesian equilibrium in signaling games? Be sure to mention the importance of the firm's beliefs, states of nature, and the signaling strategy used.

24.5 Consider the capital investment game depicted in Fig. 24.7. Suppose that $\theta = 0.5$, $Pr(\pi_H) = 0.7$, and $Pr(\pi_H | \theta) = 0.1$.

a. Use Bayes' theory to calculate conditional expected payoffs. Illustrate the extensive form of this game.
b. What is the subgame-perfect Bayesian equilibrium from a pooling strategy?

c. What is the subgame-perfect Bayesian equilibrium from a separating strategy?
d. Compare your answers to parts b and c. What, if anything, can you say about the existence of a unique perfect Bayesian equilibrium in signaling games?

24.6 Suppose that the cost of attending college for a high-productivity type is $2. The cost of attending college for a low-productivity type is $4. The value of the marginal product of a high-productivity type is $15. The value of the marginal product of a low-productivity type is $10. The population distribution of high-productivity and low-productivity types is 25 percent and 75 percent, respectively. The probability that a college graduate is high-productivity is 60 percent. The probability that a college graduate is low-productivity is 40 percent.
a. What is the conditional probability that a high-productivity type attends college?
b. What is the conditional probability that a low-productivity type attends college?
c. What is the firm's conditional expected payoff if it offers a wage of $7 to a college graduate?
d. Suppose that the next best alternative wage of high-productivity and low-productivity types is 65 percent of the value of their marginal products. What is the separating equilibrium if the firm offers a wage of $7?
e. What is the separating equilibrium if the firm offers a wage of $8?
f. What is the separating equilibrium if the firm offers a wage of $6?

24.7 Suppose that firm A must decide whether or not to launch a new product. Firm A knows whether its new product is weak (W) or strong (S). Firm B does not know whether firm A has a weak or a strong product and must also decide whether to launch or not launch its own new product. If A launches a strong product and B launches, B's profit is –200 and A's profit is 600. If A launches a strong product and B does not launch, A's profit is 800 and B's profit is zero. If A launches a weak product and B launches, B's profit is 200 and A's profit is –100. If A launches a weak product and B does not launch then A's profit is 200. Suppose that firm B believes that there is a 40 percent probability that A has a strong product and a 60 percent probability that A has a weak product. Since it is in A's best interest if B does not launch, suppose that firm A were to "leak" the results of test data indicating that the product is strong. The problem for firm B is that it does not know for certain whether the results of the test data are true, or whether the results have been falsified to make a weak product appear to be strong. In other words, firm B does not know if firm A is "bluffing." Suppose that firm B believes that there is a 25 percent chance that the report is accurate and a 75 percent chance that the report is a bluff.

a. What is the Bayesian subgame-perfect equilibrium if firm B adopts a pooling strategy?
b. What is the Bayesian subgame-perfect equilibrium if firm B adopts a separating strategy? Show your work.

Notes

1. Largely on the basis of his pioneering work on signaling, Michael Spence was awarded the 1981 John Bates Clark Medal by the American Economic Association. First awarded in 1947 to Paul Samuelson, the John Bates Clark Medal is awarded every other year to the most promising economist less than forty years of age. Michael Spence was further recognized for his work in signaling, which spawned an enormous growth in the literature on contract theory, when he shared the 2001 Nobel Prize in economics with George Akerlof and Joseph Stiglitz.

2. Other examples of signals in the marketplace include seller warranties and guarantees. The warranty on a high-quality product that is not expected to fail is likely to be longer and more generous since the expected cost to the seller is small. Other signals frequently encountered include the use of brand name, image, and reputation by manufacturers to signal uninformed buyers about the quality of the product.

CHAPTER TWENTY-FIVE

Screening

Introduction

In our discussion of signaling games in the previous chapter, the player with private information moved before a player with imperfect information. In the Spence education game, for example, a job seeker uses observable education as a signal for her unobservable productivity. In this chapter, we will examine a related phenomenon known as **screening**. In screening games, the reverse is true. The player with private information moves *after* the player with imperfect information.

When an individual applies for a mortgage loan, for example, the bank or mortgage broker requests copious information about the applicant's salary, income tax returns, checking and savings account balances, other financial assets, real assets (such as automobiles and real estate), credit history, and employment history. If the prospective borrower is a business, the lender will require income statements, balance sheets, feasibility studies, and other supporting documentation. The lending officer then screens good from bad credit risks when deciding whether or not to extend the loan. Screening is also used by prospective employers to sort applicants according to their productivity characteristics.

Self-Selection

One type of screening mechanism that is frequently encountered in the labor market is **self-selection**. To screen out unqualified candidates, applicants may be told that there are prerequisites to employment, such as a certain level of education, minimum grade point average, specific work experience, licenses, and professional certifications. Only those candidates who satisfy these requirements may continue the application process.

Prospective employers may also attempt to sort candidates according to their talents. Suppose, for example, that Ben has the option of applying for a management position that pays a salary of \$50,000 or a sales position that pays \$40,000. Ben knows that he may not have what it takes to be an effective manager and would only be able to generate \$30,000 in value added to the firm. On the other hand, Ben is a self-starter with excellent interpersonal skills. As a salesperson, Ben could add \$60,000 of value to the firm. Although Ben knows that he would make a better salesperson, he prefers the higher salary of a manager. Unfortunately, the firm's recruitment officer does not have enough information about Ben to know what is the best fit.

One possible solution would be to offer Ben the choice of either job at a guaranteed annual salary of \$20,000 annually plus a bonus of \$5,000 for each \$10,000 of value added to the firm. This compensation scheme makes it possible for Ben to earn \$20,000 + \$5,000(\$60,000/\$10,000) = \$50,000 as a salesperson, compared with only \$20,000 + \$5,000(\$30,000/\$10,000) = \$35,000 as a manager. In this way, self-selection by Ben results in an outcome that benefits both players.

Spence Education Game in Reverse

Consider the following variation of the Spence education game in which a firm hires only college graduates. In this screening game, which is depicted in Fig. 25.1, there are two types of workers with college degrees: high-productivity (H) and low-productivity (L). Since only the worker knows his or her productivity type, the firm would like to screen out low-productivity types through the process of self-selection. Applying a Harsanyi transformation, Nature moves first by determining the probability distribution of productivity types with college degrees, where $Pr(C \mid H)$ is the probability of a high-productivity college graduate and $1 - Pr(C \mid H) = Pr(C \mid L)$ is the probability of a low-productivity college graduate. The firm (F) makes an offer (O) at a specified wage (w) to a prospective worker (W) who either accepts (A) or rejects (R) the offer.

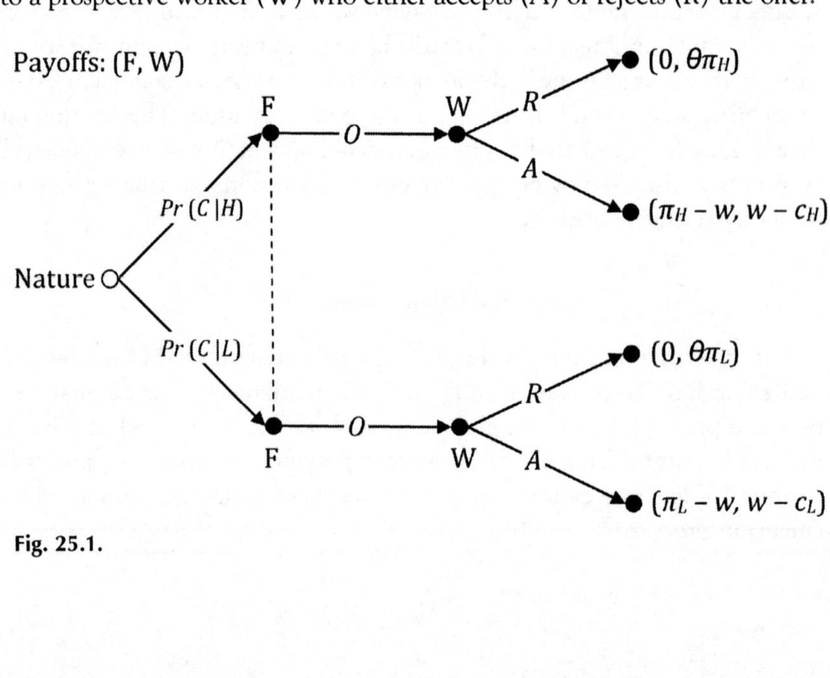

Fig. 25.1.

The value of the marginal product of high-productivity and low-productivity types is π_H and π_L, respectively. The applicant's next best alternative is self-employment in which high-productivity types earn $\theta\pi_H$ and low productivity types earn $\theta\pi_L$, where $0 < \theta < 1$. As before, the opportunity cost of attending college is greater for low-productivity types (c_L) than for high-productivity types (c_H). The payoff to the firm from hiring a high-productivity type is $\pi_H - w$ and the payoff from hiring a low-productivity type is $\pi_L - w$. The payoff to a high-productivity type is (the present value of) $w - c_H$. The payoff to a low-productivity type is $w - c_L$. The candidate will accept the job offer whenever the offered wage less the cost of college is greater than the earnings from self-employment, otherwise the offer will be rejected.

As an example, suppose that the value of the marginal product of high-productivity and low-productivity types are π_H = \$12 and π_L = \$8, respectively. The cost of attending college for high-productivity and low-productivity types are c_H = \$1 and c_L = \$6, respectively. Thus, the payoff to the firm from a high-productivity worker is $\pi_H - w$ = \$12 – \$9 = \$3 and the payoff from low-productivity workers is $\pi_L - w$ = \$8 – \$9 = –\$1. If $\theta = 0.5$ and the probability distribution of high-productivity college graduates is *Pr* (*H* | C) = 0.8 and *Pr* (*L* | C) = 0.2 for low-productivity college graduates, the expected payoffs for this screening game are summarized in Fig. 25.2.

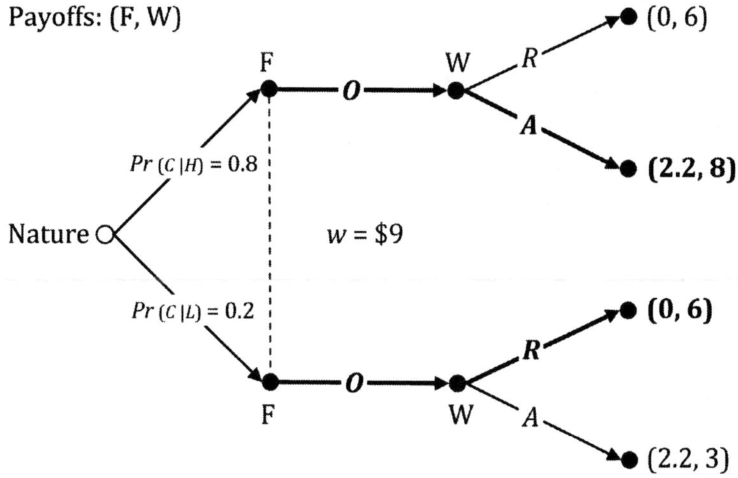

Fig. 25.2.

The subgame-perfect Bayesian equilibrium for this screening game is for the firm to make a wage offer and for high-productivity types to accept and for low productivity workers to reject. Given the size of the firm's wage offer, it does not pay for low-productivity types to attend college. In this screening game, it is not necessary for the firm to update its beliefs since it does not observe worker education signals prior to making an offer.

Solved Exercise

Consider the game depicted in Fig. 25.2.

a. *What is the subgame-perfect Bayesian equilibrium if the firm makes an offer of w = \$6?*
b. *What is the subgame-perfect Bayesian equilibrium if the firm makes an offer of w = \$12?*

Solution

a. *For high-productivity types to accept and for low-productivity types to reject, the firm's wage offer must satisfy the condition*

$$\theta\pi_L + c_L > w > \theta\pi_H + c_H. \tag{25.1}$$

Substituting the given values into condition (25.1), the firm must make an offer of

$$0.5(\$8) + \$6 = 10 > w > \$7 = 0.5(\$12) + \$1. \tag{25.2}$$

Both productivity types will reject an offer of w = \$6.

b. *Both productivity types will accept an offer of w = \$12.*

Practice Exercises

25.1 What is the strategic difference between signaling and screening games?

25.2 Why is it necessary to calculate posterior probabilities in signaling games, but not in screening games?

25.3 Consider the game depicted in Fig. 25.2. What is the subgame-perfect Bayesian equilibrium for this game when $\theta = 0.75$? What is the subgame-perfect Bayesian equilibrium if $\theta = 0.25$?

~

Appendices

Appendix A

For an infinitely repeated static game, Eq. (6.3) may be rewritten as

$$PV = \pi\left[1 + \frac{1}{(1+i)^1} + \frac{1}{(1+i)^2} + \cdots\right] = \pi S, \quad \text{(A.1)}$$

where

$$S = 1 + \frac{1}{(1+i)^1} + \frac{1}{(1+i)^2} + \cdots. \quad \text{(A.2)}$$

Multiplying both sides of Eq. (A.2) by $1/(1+i)$ we obtain

$$S\left(\frac{1}{1+i}\right) = \frac{1}{(1+i)^1} + \frac{1}{(1+i)^2} + \frac{1}{(1+i)^3} + \cdots. \quad \text{(A.3)}$$

Subtracting Eq. (A.3) from Eq. (A.2) yields

$$S - S\left(\frac{1}{1+i}\right) = 1. \quad \text{(A.4)}$$

Solving Eq. (A.4) for S we obtain

$$S = \frac{1+i}{i}. \quad \text{(A.5)}$$

Finally, substituting this result into Eq. (A.1) we get

$$PV = \pi\left(\frac{1+i}{i}\right). \tag{A.6}$$

Appendix B

For an infinitely-repeated static game, Eq. (7.3) may be written

$$PV^j = \pi^j\left[\left(\frac{1-p}{1+i}\right)^0 + \left(\frac{1-p}{1+i}\right)^1 + \left(\frac{1-p}{1+i}\right)^2 + \cdots\right], \tag{B.1}$$

which may be rewritten as

$$PV^j = \pi^j S, \tag{B.2}$$

where

$$S = 1 + \left(\frac{1-p}{1+i}\right)^1 + \left(\frac{1-p}{1+i}\right)^2 + \left(\frac{1-p}{1+i}\right)^3 + \cdots. \tag{B.3}$$

Multiplying both sides of Eq. (B.3) by $(1 - p)/(1 + i)$ we get

$$S\left(\frac{1-p}{1+i}\right) = \left(\frac{1-p}{1+i}\right) + \left(\frac{1-p}{1+i}\right)^2 + \left(\frac{1-p}{1+i}\right)^3 + \cdots. \tag{B.4}$$

Subtracting Eq. (B.4) from Eq. (B.3) we get

$$S - S\left(\frac{1-p}{1+i}\right) = 1. \tag{B.5}$$

Solving for S we get

$$S = \left(\frac{1+i}{i+p}\right). \tag{B.6}$$

Substituting Eq. (B.6) into Eq. (B.2) we get

$$PV^j = \pi^j\left(\frac{1+i}{i+p}\right). \tag{B.7}$$

Appendix C

Expected Value

The most commonly used summary measure of random payoffs is the *mean*, which is also known as the **expected value**. Random payoffs may refer to profits, capital gains, prices, unit sales, and so on. The mean is a weighted average of all possible random outcomes (x_i), with the weights being the probabilities (p_i) of each random outcome. More formally, the expected value of a discrete random outcome is

$$E(x) = \mu = p_1x_1 + p_2x_2 + \cdots + p_nx_n = \sum_{i=1}^{n} p_ix_i, \tag{C.1}$$

where $p_1 + p_2 + \ldots + p_n = 1$. When the probability of each outcome is the same, the expected value is the sum of the outcomes divided by the number of observations, or

$$E(x) = \mu = \left(\frac{1}{n}\right)\sum_{i=1}^{n} x_i. \tag{C.2}$$

Variance

By itself, the expected value tells us nothing about the riskiness of a decision. The outcomes of decisions may have the same expected payoffs, but they may not have the same risk. As we will soon discover, if two investments have the same expected value, but the first is less risky than the second, the second will most likely be preferred.

To illustrate this last point, suppose that a manager must decide whether to invest in project A or project B. The payoff from either investment depends on two states of nature—economic prosperity or recession—which are equally likely to occur. Suppose that in a prosperous economy the actual payoff from project A is \$101, while the payoff in a recession is \$99. The expected payoff from this investment is 0.5(\$101) + 0.5(\$99) = \$100. By contrast, the actual payoff from an investment in project B in a prosperous economy is \$200, but \$0 during a recession. Once again, the expected payoff from this investment is 0.5(\$200) + 0.5(\$0) = \$100. Even though the two projects have the same expected payoff, project B is riskier than project A. While it is true that the payoff from project B is so much greater than project A if the economy is prosperous, the manager runs the risk of losing his or her entire investment if the economy goes into a recession. By contrast, the worst that the manager can do by investing in project A is to lose just \$1.

It should be clear from the above example that risk is defined in terms of the volatility of payoffs. A common measure of this volatility is the **variance**, which is defined as the sum of the squared deviations of each payoff from its mean multiplied by probability of its occurrence. The variance is given by the equation

$$\sigma^2 = E[x_i - \mu]^2 = \sum_{i=1}^{n} [x_i - \mu]^2 p_i. \tag{C.3}$$

When the probability of each outcome is the same, the variance is the sum of the squared deviations of each payoff from its mean divided by the number of outcomes, which is given by the equation

$$\sigma^2 = E[x_i - \mu]^2 = \frac{1}{n}\sum_{i=1}^{n} [x_i - \mu]^2. \tag{C.4}$$

Returning to the above example, the variance for project A is σ_A^2 = [(\$101 – \$100)^2 + (\$99 – \$100)^2]/2 = 1 and the variance for project B is σ_B^2 = [(\$200 – \$100)^2 + (0 – \$100)^2]/2 = \$10,000. Based on these calculations of the variance, project B is clearly riskier than project A.

Standard Deviation

An alternative measure of risk is the **standard deviation**, which is

$$\sigma = \sqrt{\sigma^2}. \tag{C.5}$$

The standard deviation is generally preferred to the variance because it is expressed in values that are closer in magnitude to the original data. To see this, note that the mean for both projects in the above example was \$100, yet the variance for project B was \$10,000! By contrast, the standard deviations for the payoffs from the two projects were σ_A = 1 and σ_B = \$100, which in both instances is exactly equal to the variation on either side of the mean.

Glossary

Adverse selection—an ex ante asymmetric information problem in which the market becomes crowded with individuals or products having undesirable characteristics.

Asymmetric impatience—when players in a dynamic bargaining game use the different discount rates to evaluate the present value of their share of the bargaining surplus.

Asymmetric information—when some players possess private information that is not shared by others.

Auction—a forum in which buyers and sellers compete for the right to purchase or sell something of value.

Auction environment—the characteristics of bidders in an auction.

Auction rules—the procedures that govern the bidding process.

Auctioneer—the seller's agent in an auction.

Backward induction—a solution algorithm for identifying subgame-perfect equilibrium in sequential-move games.

Bargaining—the process of negotiating the terms of a contract.

Bargaining surplus—the difference between the buyer's and the seller's reservation prices.

Bayes' theorem—a statistical relationship that enables a player to update the probability of an event after learning that some other event has occurred.

Bayesian Nash equilibrium—a Nash equilibrium in a static Bayesian game.

Bayesian updating—the process of revising conditional expectations on the basis of new information.

Belief profile—a player's probability assessments at each nontrivial information set.

Bertrand model— a static, price-setting game in which firms cannot switch strategies and industry output adjusts to market demand.

Bertrand-Nash equilibrium—a Nash equilibrium in a Bertrand price-setting game.

Bertrand paradox—the prediction not observed in the real world that price-setting duopolists producing perfect substitutes set price equal to marginal cost and earn zero economic profit.

Best-response (reaction) function—a formal statement of a player's best response to a move made by a rival.

Bluff—misdirection to gain a strategic advantage.

Branches—a player's moves in a game tree.

Breach of contract—when a party to a contract fails to perform, or refuses to perform, a specified obligation.

Breach remedy—court-ordered compensation in the event of a breach of contract.

Buyer's reservation price—the maximum price that the buyer is willing to pay.

Cartel—a formal agreement among firms in an industry to allocate market share, coordinate pricing strategies, or some other coordinated action to increase industry profits.

Chance node—a decision node where the probabilities of random events are determined.

Closed-bid auction—an auction in which participation is by invitation only.

Coalition game—a game in which it is in the best interest of the players to cooperate, but each player has an incentive to adopt a strategy that results in a less-than-optimal outcome.

Collusion—when firms coordinate their decisions to restrict competition and increase market power.

Commercial policy—regulations that define a country's international trade relations.

Common-value auction—when bidders have different private estimates of the true, but unknown, value of the object being auctioned.

Competition game—a game in which the players rank their payoffs from alternative strategy profiles in reverse order.

Complete information—the players, their strategy options, and payoffs are common knowledge.

Complete-information auction—bidders' private valuations are common knowledge.

Consumer surplus—the total value buyers receive less total expenditures.

Contract—a formal agreement that obligates the parties to perform, or refrain from performing, a specified act in exchange for something of value.

Coordination game—a game in which players coordinate strategies to maximize joint payoffs.

Correlated value estimates—when bidders' private value estimates affect, and are affected by, the private valuations of other bidders.

Cournot model—a static, output-setting game in which rival firms simultaneously determine their optimal output levels. Once determined, output cannot be changed, or at least not at significant economic cost.

Cournot-Nash equilibrium—a Nash equilibrium in a Cournot output-setting game.

Decision node—a branching point in a game tree where a player moves.

Discount rate—the interest rate used to determine the present value of future payoffs.

Dominant strategy—a strictly-dominant or a weakly-dominant strategy.

Double marginalization—occurs when quasi-independent upstream and downstream divisions of the same firm set the price of their output above marginal cost to maximize their respective profits, which comes at the expense of maximizing company profits.

Dropout price—the price at which a player drops out of an English auction.

Dutch (open-descending-bid) auction—the auctioneer incrementally lowers an initially high price that no one accepts. The first bidder to accept is declared the winner.

Dynamic (sequential-move) game—a game in which the players alternate moves.

Economics of information—the study of decision making under conditions of risk and uncertainty.

End-of-game problem—A player's move in the last stage of a finitely-repeated static game with a certain end determines how that player will move in the first and all subsequent stages.

English (open-ascending-bid) auction—the auctioneer opens the bidding with a low asking price. As prices are incrementally increased bidders who value the object less drop out. The last bidder remaining wins.

Evolutionary equilibrium—a population-dynamic strategy profile that is resistant to mutation.

Evolutionary game theory—a branch of game theory that explains animal behavior in terms of instinctual strategies that are genetically passed along to future generations.

Expected value—the weighted average of a set of random outcomes, where the weights are the probabilities of each outcome.

Export subsidy—a payment made by government to a domestic company to increase foreign sales.

Extensive form—a formal representation of a game that summarizes the players, the information available at each stage of the game, and the payoffs from alternative strategy profiles.

Fair gamble—a gamble with a zero expected payoff of zero.

Finitely-repeated static game—a static repeated game with a certain or uncertain end.

First-mover advantage—a player who can commit to a strategy first enjoys a payoff that is no worse than if all players move simultaneously.

Focal-point equilibrium—when a strategy profile emerges from among multiple Nash equilibria because the players have a shared understanding of the strategic environment.

Framing—when players make decisions based on a common frame of reference.

Game theory—the formal study of decision making in situations involving move and countermove.

Game tree—the diagram of the extensive form of a game.

Grim strategy—a trigger strategy in a repeated game in which a player never again enters into a coalition once another player has defected.

Horizontal differentiation—a change that makes a firm's product better in the eyes of some consumers, but worse in the eyes of others.

Imperfect information—games involving asymmetric information.

Improper subgame—a subgame with a subroot located within a valid information set of more than one decision node.

Incentive contract—a formal agreement that links a manager's compensation to company performance.

Incomplete information—games in which payoffs are not known with certainty.

Incomplete-information auction—bidders know their own private valuations, but not the private valuations of rivals.

Independent investments—when returns from alternative investments are unrelated and acceptance of one investment does not preclude acceptance of other investments.

Independent private values—bidders' private valuations unaffected by the private value estimates of other bidders.

Infinitely-repeated static game—a static game that is played repeatedly without end.

Information partition—a collection of information sets.

Information set—a collection of decision nodes. A player within an information set is unable to distinguish one decision node from another.

Information structure—the availability of bidders' private valuations, and whether private valuations are correlated.

Intraindustry trade—when a country exports and imports closely related goods and services.

Invalid information set—a collection of decision nodes for two or more players where a decision node by one player precedes the decision node of another player in the same information set.

Investor indifference curve—a diagram that summarizes the combinations of risk and expected returns for which an investor is indifferent.

Last-mover advantage—when the player who makes the last counteroffer in a bargaining game with a finite number of negotiating rounds extracts the full amount of the bargaining surplus.

Marginal utility of money—the additional satisfaction received from an additional dollar of money income or wealth.

Market failure—when the forces of supply and demand fail to generate socially efficient levels of consumption and production.

Maximin (secure) strategy—a strategy maximizes a player's minimum payoffs.

Minimax theorem—in a two-player, zero-sum game there exists a mixed strategy equilibrium in which the players maximize their minimum (maximin) expected gain and minimize their maximum (minimax) expected loss.

Mixed strategy—randomly mixing pure strategies.

Mixed-strategy Nash equilibrium—a unique Nash equilibrium that results when all players adopt their optimal mixing rules.

Moral hazard—an ex post asymmetric information problem in which there is the risk (hazard) that a party to a transaction engages in activities that are undesirable (immoral) from the perspective of the other party to the same transaction.

Mutually exclusive investments—when acceptance of one investment precludes acceptance of all other investments.

Nash bargaining—a static game in which the players submit bids for a share of a divisible object of known value. If the sum of the bids is less than the amount available, the players receive their bids. If the sum of the bids is greater than the amount available, the players receive nothing and resubmit a revised bid.

Nash equilibrium—when each player adopts a strategy that is the best response to the strategies adopted by the other players, and neither player can get a better payoff by unilaterally switching strategies.

Nash equilibrium existence theorem—randomly mixing pure strategies using an optimal mixing rule is sufficient to guarantee that every game with a finite number of players and pure strategies has a unique Nash equilibrium.

Network—a system of integrated connections involving a common technical platform.

Non-zero-sum game—when the sum of the players' payoffs does not equal zero.

Normal form—a formal representation of a static game that summarizes the players, strategies, and payoffs from alternative strategy profiles.

Oligopoly—an industry dominated by a few large firms producing perfect or near perfect substitutes in which pricing and output decisions are interdependent.

One-time game—a game that is played just once.

Open-bid auction—an auction in which anyone may submit a bid.

Optimal mixing rule—a randomized mix of pure strategies in which players maximize their minimum expected gains and minimize their maximum expected losses.

Oral auction—bids are announced publicly.

Payoff—the expected gain or loss that results from a player's strategy choices.

Perfect Bayesian equilibrium—a subgame perfect equilibrium that is consistent with the probability distribution of states of nature and players' beliefs.

Perfect information—when players know their precise location in a game tree.

Player—a decision maker in a game.

Player type—a player's personal characteristics, tastes, and preferences.

Pooling equilibrium—a subgame-perfect Bayesian equilibrium that results from a pooling strategy.

Pooling strategy—when the same strategy is adopted for different player types. A pooling strategy does not use signals to distinguish player types.

Positive network externalities—when the value of a individual network membership increases exponentially with the number of users.

Predatory (limit) pricing—the practice of a charging price that is low enough to deter entry into the market by potential rivals.

Price leadership—a form of price collusion in which a dominant firm initiates a price change that is matched by the rest of the industry.

Principal-agent problem—when one group of stakeholders called agents do not act in the best interests of another group of stakeholders called principals because their interests are not properly aligned.

Procurement (reverse) auction—multiple sellers compete for the right to sell something of value.

Producer surplus—a firm's operating profit.

Promise—a countermove that rewards a rival for cooperating.

Proper subgame—a subgame that begins at a decision node other than the root of the game tree.

Pure strategy—a complete and nonrandom game plan.

Rational behavior—when players make choices that are in their self-interest.

Reciprocal (two-way) network—a network in which data flows in two directions.

Repeated game—a game that is played multiple times.

Reproductive success—a weaker assumption than pure rationality in which successful strategies replace unsuccessful strategies over time.

Risk—multiple payoffs from alternative states of nature with known probabilities.

Risk averse—when a person prefers a certain payoff to a risky prospect with the same expected value.

Risk loving—when a person prefers a risky prospect of an expected payoff to its certainty equivalent.

Risk neutral—when a person is indifferent between a certain payoff and a risky prospect with the same expected value.

Root of a game tree—the first decision node in a game tree.

Rubenstein bargaining—sequential-move bargaining involving a potentially infinite number of offers and counteroffers.

Screening—when an uninformed individual attempts to sort players according to defined characteristics.

Sealed-bid auction—bids are submitted secretly.

Sealed-bid, first-price auction—the winner submits, and pays, the highest sealed bid.

Sealed-bid, second-price (Vickery) auction—the winner submits the highest sealed bid, but pays the price submitted by the second-highest bidder.

Self-selection—when players sort themselves into different groups according to predefined characteristics.

Seller's reservation price—the minimum price that a seller is willing to accept.

Separating equilibrium—a subgame-perfect Bayesian equilibrium that results from a separating strategy.

Separating strategy—when a different strategy is adopted for different player types. A separating strategy uses signals to distinguish player types.

Signaling—when a player transmits a credible and unreproducible signal that conveys information that is not otherwise observable to other players.

Simultaneous-move game—a game in which all players move at the same time.

Stackelberg follower—a firm that commits to a strategy after observing the output of the Stackelberg leader.

Stackelberg leader—a firm that commits to a strategy first in an output-setting game.

Stackelberg model—an output-setting, dynamic game in which the Stackelberg leader exploits the best-response function of the Stackelberg follower to maximize profits.

Standard auction—multiple buyers compete for the right to purchase something of value.

Standard deviation—the square root of the variance. The standard deviation is a commonly used measure of the riskiness of an investment.

Static Bayesian game—a static game with incomplete information.

Static game—a game in which the players do not know each other's moves until all moves have been made.

Strategic behavior—when the decisions of individuals and groups affect, and are affected by, the decisions of other individuals and groups.

Strategic complements—when the decisions of individual players are mutually reinforcing.

Strategic move—an observable and irreversible move designed to alter a rival's behavior.

Strategic trade policy—policies intended to influence the strategic relationship between countries engaged in international trade.

Strategy—a game plan for achieving a player's objective.

Strategy profile—the collection of all players' strategies.

Strictly-dominant strategy—the same best response to any strategy adopted by another player.

Strictly-dominated strategy—a strategy that is dominated by every other strategy.

Subgame—a subset of branches and decision nodes in a game tree.

Subgame equilibrium—the equilibrium of a proper subgame.

Subgame-perfect equilibrium—the equilibrium of a trivial subgame, which is found among the complete collection of subgame equilibria.

Subroot—the root of a subgame.

Symmetric impatience—when players in a dynamic bargaining game use the same discount rate to evaluate the present value of their share of the bargaining surplus.

Terminal node—the point at which the game ends.

Threat—a countermove intended to punish rivals who do not cooperate.

Tit-for-tat—a strategy that mimics a rival's behavior.

Total (perfect) recall—when players remember everything that happens during a game, including earlier moves made by other players.

Trigger strategy—a move made in response to an unanticipated move by another player.

Trivial subgame—a subgame that begins at the root of the game tree.

Ultimatum game—a take-it-or-leave-it bargaining scenario.

Uncertainty—multiple payoffs from alternative states of nature with unknown or meaningless probabilities.

Unidirectional (one-way) network—a network in which data flow in one direction.

Valid information set—a collection of decision nodes for the same player where no decision node precedes any other decision node in the same information set.

Variance—a measure of the dispersion of a set of random outcomes. It is the sum of the products of the squared deviations of each outcome from its mean and the probability of each outcome.

Vertical differentiation—a change that makes a firm's product better in the eyes of everyone.

Virtual network—a network involving compatible goods and services sharing a common technical platform

Weakly-dominant strategy—a strategy that results in the same payoff in response to one strategy and a strictly better payoff in response to another strategy.

Winner's curse—when the bidder with the most optimistic private value estimate wins a common value auction, but pays a price that is greater than the object's true value.

Zero-sum game—when one player's gain is another player's loss.

References and Suggestions for Further Reading

Acheson, J. M. 1975. The lobster fiefs: Economic and ecological effects of territoriality in the maine lobster industry. *Human Ecology* 3:183–207.

Akerlof, G. 1970. The market for 'lemons': Quality, uncertainty, and the market mechanism. *Quarterly Journal of Economics* 84: 488–500.

Arrow, K. 1970. *Essays in the Theory of Risk Bearing*. Chicago: Markham.

Aumann, R. 1987. Correlated equilibrium as an extension of bayesian rationality. *Econometrica*, 55(1):1–18.

Axelrod, R. 1984. *The Evolution of Cooperation*. New York: Basic Books.

Baghwati J., V. Ramaswami, and T. Srinivasam. 1969. Domestic distortions, tariffs and the theory of optimum subsidy: Some further results. *Journal of Political Economy* 77:1005–10.

Bain, J. S. 1956. *Barriers to New Competition*. Cambridge, MA: Harvard University Press.

Baird, D. G., R. H. Gertner, and R. C. Picker. 1994. *Game Theory and the Law*. Cambridge, MA: Harvard University Press.

Bajari, P. and A. Hortacsu. 2004. Economic insights from internet auctions. *Journal of Economic Literature* 42(2):457–486.

Barry, S., J. Levinsohn, and A. Pakes. 1995. Voluntary export restraints on automobiles: Evaluating a strategic trade policy. *National Bureau of Economic Research Working Paper No. 5235* (August).

———. 1995. Voluntary export restraints in automobiles. *American Economic Review* 89(3):400–430.

Bastiat, F. 1848. *Selected Essays on Political Economy*, trans. Seymour Cain, ed. George B. de Huszar, introduction by F.A. Hayek. Irvington-on-Hudson: Foundation for Economic Education (1995).

Baumol, W. J., J. C. Panzar, and R. D. Willig. 1982. *Contestable Markets and the Theory of Industry Structure*. New York: Harcourt Brace Jovanovich.

Bazerman, M. and W. Samuelson. 1983. I won the auction but don't want the prize. *Journal of Conflict Resolution* 27(4):618–34.

———. 1983. The winner's curse: an empirical investigation. In *Aspiration Levels in Bargaining and Economic Decision Making*, ed. Reinhard Tietz, 186–200. New York: Springer-Verlag.

Behnke, R. 1985. *Open Range Management and Property Rights in Pastoral Africa: A Case for Spontaneous Range Enclosure in South Darfur, Sudan*. Pastoral Development Network Paper. London: Overseas Development Institute.

Berg, A. S. 1989. *Goldwyn: A Biography*. New York: Alfred A. Knopf.

Bernoulli, D. 1738. Specimen theoriae novae de mensura sortis. *Commentarii Academiae Scientiarum Imperialis Petropolitannae*. Translated by L. Sommer as: Exposition of a new theory on the measurement of risk, *Econometrica* 22, (1954):23–36.

Bertrand, J. 1883. Théorie mathématique de la richesse sociale. *Journal des Savants* 67:499–508.

Binmore, K. 1992. *Fun and Games: A Text on Game Theory*. Lexington, MA: D. C. Heath.

Binmore, K. and P. Dasgupta. 1986. *Economic Organizations as Games*. Oxford: Basil Blackwell.

Blomquist, W. and E. Ostrom. 1985. Institutional capacity and the resolution of the commons dilemma. *Policy Studies Review* 5:383–393.

Blount, S. 1995. When social outcomes aren't fair: the effect of casual attributions on preferences. *Organizational Behavior and Human Decision Processes* 63(2):131–144.

Borel, E. 1921. La theorie du jeu et les equations integrales a noyen symetrique gauche. *Comptes Rendus Hebdomadaires des Seances de l'Academie Sciences* 173:1304–1308. Translated by Leonard J. Savage as: The theory of play and integral equations with skew symmetric kernels, *Econometrica* 21(1953):97–100.

———. 1923. Sur les jeuxs ou interviennent l'hasard et l'habilete des joueurs. *Association Francaise pour l'Advancement des Sciences*, 79–85.

———. 1924. *Elements de la Theorie des Probabilites*. 3d ed. Paris: Librairie Scientifique, J. Hermann, 204–221. Translated by Leonard J. Savage as: On games that involve chance and the skill of the players, *Econometrica* 21(1953):101–115.

———. 1924. Sur les jeux ou interviennent l'hasard et l'habilete des joueurs. *Theorie des Probabilites*. Paris: Librairie Scientifique, J. Hermann, 204–224.

———. 1926. Un theoreme sur les systemes de formes lineaires a determinant symetrique gauche. *Comptes Rendus Academie des Sciences* 183:925–927, Avec Erratum: 996.

———. 1927. Sur les systemes de formes lineaires a determinant symetrique gauche et la theorie generale du jeu. *Comptes Rendus Hebdomadaires des Seances de l'Academie de Sciences* 184: 52–53. Translated by Leonard J. Savage as: On systems of linear forms of skew symmetric determinant and the general theory of play, *Econometrica* 21(1953):116–117.

———. 1938. Traite du calcul des probabilites et de ses applications. *Applications des jeux de hasard* (Lectures by Emile Borel, transcribed by Jean Ville). Paris: Gauthier-Villars, Tome. IV, Fascicule 2.

———. 1938. Jeux ou la psychologie joue un role fondamental. *Applications Des Jeux de Hasard* (Lectures by Emile Borel, transcribed by Jean Ville). Paris: Gauthier-Villars, Tome. IV, Fascicule 2.

Brams, S. and M. Kilgour. 1988. *Game Theory and National Security*. Oxford: Basil Blackwell.

Brander, J. 1981. Intra-industry trade in identical commodities. *Journal of International Economics* 11:1–14. Reprinted in Intra-Industry Trade, ed. P. Lloyd and H. Grubel. Camberly, UK: Edward Elgar, 2002.

Brander, J. A. and B. J. Spencer. 1981. Tariffs and the extraction of foreign monopoly rent under potential entry. *Canadian Journal of Economics* 14:371–89. Reprinted in *Readings in International Trade*, ed. J. Bhagwati. Cambridge, MA: MIT Press, 1987.

———. 1983. International R&D rivalry and industrial strategy, *Review of Economic Studies* 50:707–722.

———. 1985. Export subsidies and international market share rivalry. *Journal of International Economics* 18: 83–100. Reprinted in *International Trade, Volume 1. Welfare and Trade Policy*, ed. P. Neary, Aldershot, UK and Brookfield, VT: Edward Elgar, 1995.

Bulow, J. I., J. D. Geanakoplos, and P. D. Klemperer. 1985. Multimarket Oligopoly: Strategic Substitutes and Strategic Complements. *Journal of Political Economy* 93(3): 488–511.

Camerer, C. F., and R. H. Thaler. 1995. Anomalies: Ultimatums, dictators and manners. *Journal of Economic Perspectives* 7(2):159–171.

———. 2003. In honor of Matthew Rabin: Winner of the John Bates Clark Medal. *Journal of Economic Perspectives* 17(3):159–176.

Camerer, C. F. 2003. *Behavioral Game Theory: Experiments in Strategic Behavior*. Princeton University Press, Princeton.

Carter, J. R., and M. D. Irons. 1991. Are economists different, and if so, why? *Journal of Economic Perspectives* 5(2):171–177.

Chamberlin, E. H. 1933. *The Theory of Monopolistic Competition*. Cambridge, MA: Harvard University Press.

Chiappori, P. A., S. Levitt, and T. Groseclose. 2002. Testing mixed-strategy equilibria when players are heterogeneous: The case of penalty kicks in soccer. *American Economic Review* 92(4):1138–1151.

Coase, R. H. 1960. The problem of social costs. *Journal of Law and Economics* 3:1–44.

Commack, E. B. 1991. Evidence on bidding strategies and the information in treasury bill auctions. *Journal of Political Economy* 99(1):100–130.

Cooper, R., D. V. DeJong, R. Forsythe, and T. W. Ross. 1996. Cooperation without reputation: Experimental evidence from prisoner's dilemma games. *Games and Economic Behavior* 12:187–218.

Cournot, A. 1838. *Recherches Sur Les Principes Mathematiques de la Theorie Des Richesses. Paris: Hachette*. Translated by Nathaniel T. Bacon as: *Researches into the Mathematical Principles of the Theory of Wealth*. New York: Macmillan, 1897.

Darwin, C. 1859. *On the Origin of Species by Means of Natural Selection*. London: John Murray.

Davis, D. D. and C. A. Holt. 1993. *Experimental Economics*. Princeton: Princeton University Press.

Davis, O. and A. Whinston. 1962. Externalities, welfare, and the theory of games. *Journal of Political Economy* 70:241–262.

de Fraja, G. and F. Delbono. 1990. Game Theoretic Models of Mixed Oligopoly. *Economic Surveys* 4:1–17.

Dimand, R. W. and M. Dimand. 1992. The early history of the theory of strategic games from Waldegrave to Borel. In *Toward a History of Game Theory*, ed. E. R. Weintraub. Durham: Duke University Press.

Dixit, A. 1979. A model of duopoly suggesting a theory of entry-barriers. *Bell Journal of Economics* 10:20–32.

———. 1984. International trade policy for oligopolistic industries. *Economic Journal* 94 (supplement):1–16.

Dixit, A. and B. Nalebuff. 1991. *Thinking Strategically: The Competitive Edge in Business, Politics, and Everyday Life*. New York: W. W. Norton.

Dixit, A. and S. Skeath. 1999. *Games and Strategy*. New York: W. W. Norton.

Doyle, A. C. 1890. *The Sign of the Four*. London: Lippencott's Monthly Magazine.

Dybvig, P. and J. Zender. 1991. Capital structure and dividend irrelevance with asymmetric information. *Review of Financial Studies* 4:201–219.

Eaton, J. and G. Grossman. 1986. Optimal trade and industrial policy under oligopoly. *Quarterly Journal of Economics* 101:383–406. Reprinted in *Readings in International Trade*, ed. J. Bhagwati. Cambridge, MA: MIT Press, 1987.

Economides, N. 2006. Competition policy in network industries: An introduction. In *The New Economy and Beyond: Past, Present and Future*, ed. D. W. Jansen. Northampton, MA: Edward Elgar Publishing.

Economides, N., G. Lopomo, and G. Woroch. 1996a. Regulatory pricing policies to neutralize network dominance. *Industrial and Corporate Change* 5(4):1013–28.

———. 1996b. *Strategic Commitments and the Principle of Reciprocity in Interconnection Pricing*. Discussion paper EC-96-13, Stern School of Business, New York University. Available at http://www.stern.nyu.edu/networks/96-13.edf.

Frank, R. H., T. Gilovich, and D. T. Regan. 1993. Does studying economics inhibit cooperation? *Journal of Economic Perspectives* 7(2):159–171.

Fudenberg, D. and D. K. Levine. 1998. *The Theory of Learning in Games*. Cambridge, MA: MIT Press.

Fuller, J. G. 1962. *The Gentlemen Conspirators: The Story of the Price-Fixers in the Electrical Industry*. New York: Grove Press.

Gibbons, R. 1992. *Game Theory for Applied Economists*. Princeton: Princeton University Press.

Grossman, S. J. and O. D. Hart. 1983. An analysis of the principal-agent problem. *Econometrica*, 51(1):7–45.

Güth, W., R. Schmittberger, and B. Schwarze. 1982. An experimental analysis of ultimatum bargaining. *Journal of Economic Behavior & Organization* 3(4):367–388.

Hagenmayer, S. J. 1995. Albert W. Tucker, 89, famed mathematician, *Philadelphia Inquirer* (February 2): B7.

Hansen, R. G. and W. F. Samuelson. 1988. Evolution in economic games. *Journal of Economic Behavior and Organization* 10(3):315–38.

Hardin, G. 1968. Tragedy of the commons. *Science* 162(3859) (December 13):1243–1248.

Hardin, G. and J. Baden (eds.). 1977. *Managing the Commons*. San Francisco: Freeman.

Harsanyi, J. C. 1968. Games with incomplete information played by 'Bayesian' players. Parts I–III. *Management Science* 14(3):159–182, 320–334, 486–502.

———. 1977. *Rational Behavior and Bargaining Equilibrium in Games and Social Situations*. London and New York: Cambridge University Press.

———. 1995. Games with incomplete information. *American Economic Review* 85(3):291–303.

Harsanyi, J. C. and R. Selten. 1972. A generalized Nash solution for two-person bargaining games with incomplete information. *Management Science* 18(5):80–106.

Heckscher, E. 1919. The effects of foreign trade on the distribution of income. Originally published in *Readings in the Theory of International Trade*, ed. Howard S. Ellis and Lloyd M. Metzler. Philadelphia: Blackiston, 1949.

Henrich, J., R. Boyd, S. Bowles, C. Camerer, E. Fehr, and H. Gintis (eds.). 2004. *Foundations of Human Sociality: Economic Experiments and Ethnographic Evidence from Fifteen Small-Scale Societies*. Oxford University Press, New York.

Heywood, J. S. and X. Wei. 2004. Education and Signaling: Evidence from a Highly Competitive Labor Market. *Education Economics* 12(1):1–16.

Hill, E. and D. F. Sally. 2002. *Dilemmas and Bargains: Theory of Mind, Cooperation and Fairness*. Available at SSRN: http://ssrn.com/abstract=407040 or DOI: 10.2139/ssrn.407040.

Hofbauer, J. and K. Sigmund. 1998. *Evolutionary Games and Population Dynamics*. Cambridge: Cambridge University Press.

Horowitz, I. 1991. On the effects of Cournot rivalry between entrepreneurial and cooperative firms. *Journal of Comparative Economics* 15(March):115–21.

Ivanova-Stenzel, R. and T. C. Salmon. 2004. Bidder preferences among auction institutions. *Economic Inquiry* 42(2):223–236.

Kahneman, D., J. L. Knetsch, and R. H. Thaler. 1986. Fairness and the assumptions of economics. *Journal of Business* 59(4):S285–S300.

Knight, F. 1921. *Risk, Uncertainty and Profit*. New York: Houghton Mifflin.

Kreps, D. M. 1987. Nash equilibrium. In *The New Palgrave: A Dictionary of Economics*, ed. J. Eatwell, M. Milgate, and P. Newman. London: Macmillan.

———. 1990. Corporate cultures. In J. E. Alt and K. A. Shepsle, *Perspectives in Political Economy*. Cambridge: Cambridge University Press.

Kreps, D., P. Milgrom, J. Roberts, and R. Wilson. 1982. Rational cooperation in the finitely repeated prisoners' dilemma. *Journal of Economic Theory* 27(2):245–52.

Krugman, P. 1984. Import protection as export promotion: international competition in the presence of oligopoly and economies of scale. In *Monopolistic Competition and Product Differentiation and International Trade*, ed. H. Kierzkowski. Oxford: Oxford University Press.

Lapan, H. E. and T. Sandler. 1988. To bargain or not to bargain: That is the question. *American Economic Review, Papers and Proceedings* 78(2):16–20.

Lewontin, R. C. 1961. Evolution and the theory of games. In *Toward a History of Game Theory*, ed. E. R. Weintraub. Durham: Duke University Press.

Liebowitz, S. 2006. Network meltdown: The legacy of bad economics. In *The New Economy and Beyond: Past, Present and Future*, ed. D. W. Jansen. Northampton, MA: Edward Elgar Publishing.

Lind, B. and C. Plott. 1991. The winner's curse: Expectations with buyers and with sellers. *American Economic Review* 81:225–246.

Lloyd, W. F. 1833. *Two Lectures on the Checks to Population*. Oxford: Oxford University Press. Reprinted (in part) in *Population, Evolution, and Birth Control*, ed. G. Hardin. San Francisco: Freeman, 1964, 37.

Luce, R. D. and H. Raiffa. 1957. *Games and Decisions*. New York: John Wiley & Sons.

Maynard Smith, J. 1982. *Evolution and the Theory of Games*. Cambridge: Cambridge University Press.

McAfee, R. P. and J. McMillan. 1987. Auctions and bidding. *Journal of Economic Literature* 25(2):699–783.

McDonald, J. 1950. *Strategy in Poker, Business, and War*. New York: W. W. Norton.

McMillan, J. 1994. Selling spectrum rights. *Journal of Economic Perspectives* 8(3):145–62.

———. 1992. *Games, Strategies, & Managers: How Managers Can Use Game Theory to Make Better Decisions*. London: Oxford University Press.

Milgrom, P. R. and R. J. Weber. 1982. A theory of auctions and competitive bidding. *Econometrica* 50:1089–1122.

———. 1989. Auctions and bidding. *Journal of Economic Perspectives* 3(3):3–22.

Myers, S. and N. Majluf. 1984. Corporate financing and investment decisions when firms have information that investors do not have. *Journal of Financial Economics* 13:187–221.

Myerson, R. B. 1981. Optimal auction design. *Mathematics of Operations Research* 6(1):58–73.

Myerson, R. B. and M. A. Satterthwaite. 1983. Efficient mechanisms for bilateral trading. *Journal of Economic Theory* 29:265–81.

Nash, J. 1950a. The bargaining problem. *Econometrica* 18:155–62.
———. 1950b. Equilibrium points in *n*-person games. *Proceedings of the National Academy of Sciences, USA* 36:48–49.
———. 1950c. A simple three-person poker game (with Lloyd S. Shapley). *Annals of Mathematics Study* 24.
———. 1951. Noncooperative games. *Annals of Mathematics* 51:286–295.
———. 1953a. A comparison of treatments of a duopoly situation (with J. P. Mayberry and M. Shubik). *Econometrica* 21:141–154.
———. 1953b. Two-person cooperative games. *Econometrica* 21:405–421.
Nasar, S. 1998. *A Beautiful Mind*. New York: Simon & Schuster.
Netting, R. M. 1972. Of men and meadows: Strategies of alpine and land use. *Anthropological Quarterly* 45:132–144.
———. What alpine peasants have in common: Observations on communal tenure in a Swiss village. *Human Ecology* 4:135–146.
———. Territory, property, and tenure. In *Behavioral and Social Science Research: A National Resource*, pp. 446–501. R. N. S. and D. J. Treiman Adams, eds. Washington, D.C.: National Academy Press.
Noussair, C. 1995. Equilibrium in a multi-object uniform price sealed-bid auction with multi-unit demands. *Economic Theory* 5(2):337–351.
Ohlin, B. 1933. *International and Interregional Trade*. Cambridge, MA: Harvard University Press.
Oosterbeek, H., R. Sloof, and G. van de Kuilen. 2004. Differences in the ultimatum game: Evidence from a meta-analysis. *Experimental Economics* 7(2):171–188.
Osborne, M. J. and A. Rubinstein. 1994. *A Course in Game Theory*. Cambridge, MA: MIT Press.
Ostrom, E. 1990. *Governing the Commons: The Evolution of Institutions for Collective Action*. Cambridge: Cambridge University Press.
Porter, M. 1980. General Electric vs. Westinghouse in large turbine generators. *Harvard Business School Case No. 9-380-129*.
Porter, M. and A. M. Spence. 1982. The capacity expansion decision in a growing oligopoly: The case of corn wet milling. In *The Economics of Information of Uncertainty*, ed. J. J. McCall, 259–316. Chicago: University of Chicago Press.
Poundstone, W. 1992. *Prisoners' dilemma: John von Neuman, Game Theory, and the Puzzle of the Bomb*. New York: Doubleday.
Pratt, J. W. 1964. Risk aversion in the small and in the large. *Econometrica* 32:122–136.
Pruett-Jones, S. and M. Pruett-Jones. 1994. Sexual competition and courtship disruptions: why do bowerbirds destroy each other's bowers? *Animal Behavior* 47:607–20.
Ricardo, D. 1817. *On the Principles of Political Economy and Taxation*. London: John Murray.
Riley, J. G. 1989. Expected revenues from open and sealed-bid auctions. *Journal of Economic Perspectives* 3(3):41–50.
Robert, C., and P. J. Carnevale. 1997. Group choice in ultimatum bargaining. *Organizational Behavior and Human Decision Processes* 72(2):256–279.
Rosenthal, R. W. 1982. Games of perfect information, predatory pricing, and the chain store paradox. *Journal of Economic Theory* 25:92–100.
———. 1993. Rules of thumb in games. *Journal of Economic Behavior and Organization* 22(1):1–13.
Roth, A. E., J. K. Murnighan and F. Schoumaker. 1988. The deadline effect in bargaining: some experimental evidence. *American Economic Review* 78(4):806–23.

Rubenstein, A. 1982. Perfect equilibrium in a bargaining model. *Econometrica* 61 (1):97–109.
Samuelson, L. 1998. *Evolutionary Games and Equilibrium Selection*. Cambridge, MA: MIT Press.
Schelling, T. 1960. *The Strategy of Conflict*. London: Oxford University Press.
———. 1989. Strategy and self-command. *Negotiation Journal* 5(4):343–347.
Schotter, A. 1985. *Free Market Economics: A Critical Appraisal*. New York: St. Martin's Press.
Selten, R. 1965. Spieltheoretische behandlung eines oligopolmodells mit nachfragetragheit. *Zeitschrift fur die gesamte Staatswissenschaft* 121:301–24.
———. 1975. Reexamination of the perfectness concept for equilibrium points in extensive games. *International Journal of Game Theory* 4:25–55.
Shavell, S. 1980. Damage measures for breach of contract. *Bell Journal of Economics* 11(2):466–490.
Shleifer, A. and R. W. Vishney. 1986. Greenmail, white knights, and shareholders' interest. *Rand Journal of Economics* 17(3):293–309.
Simon, H. 1957. A behavioral model of rational choice. *Quarterly Journal of Economics* 69(1):99–118.
Smith, J. M. 1982. *Evolution and the Theory of Games*. Cambridge: Cambridge University Press.
Spence, M. 1973. Job market signaling. *Quarterly Journal of Economics* 87(3):355–374.
Stahl, I. 1972. *Bargaining Theory*. Economic Research Institute, Stockholm.
Stigler, G. J. 1947. The kinky oligopoly demand curve and rigid prices. *Journal of Political Economy* 55(5):432–449.
Stolper, W. F. and Samuelson, P. A. 1941. Protection and real wages. *Review of Economic Studies* 9(1):58–73.
Sweezy, P. 1939. Demand conditions under oligopoly. *Journal of Political Economy* 47:568–573.
Tellis, G. and P. Golder. 1991. First to market, first to fail: real causes of enduring market leadership. *Sloan Management Review* 37(2):65–75.
Thaler, R. H. 1988. Anomalies: The winner's curse. *Journal of Economic Perspectives* 2(1):91–202.
Tucker, A. W. 1955. *Game Theory and Programming*. Stillwater, OK: Department of Mathematics, Oklahoma Agricultural and Mechanical College.
Vickery, W. 1961. Counterspeculation, auctions, and competitive sealed tenders. *Journal of Finance* 16:8–37.
Van't Wout, M., R. S. Kahn, A. G. Sanfey, and A. Aleman. 2006. Affective state and decision-making in the ultimatum game. *Experimental Brain Research* 169(4):564–568.
von Neumann, J. 1928. Zur theorie der gesellschaftsspiele. *Mathematische Annalen* 100:295–320. Translated as: On the theory of games of strategy. *Contributions to the Theory of Games*, IV (*Annals of Mathematics Studies*), ed. A. Tucker and R. D. Luce. Princeton: Princeton University Press, 1959.
von Neumann, J. and O. Morgenstern. 1944. *Theory of Games and Economic Behavior*. New York: John Wiley & Sons.
Von Stackleberg, H. 1934. *Marktform und Gleichgewicht*. Vienna: Julius Springer, 1934. Reprinted in *The Theory of the Market Economy*, translated by A. T. Peacock. London: William Hodge, 1952.
Walker, M. and J. Wooders. 2001. Minimax play at Wimbledon. *American Economic Review* 91(5):1521–1538.
Williams, J. D. 1966. *The Compleat Strategyst*. New York: McGraw-Hill.

Index

positive feedback effects, 66–67
positive network externalities. *See* positive feedback effects

About the Author

Thomas J. Webster is a professor of economics in the Department of Finance and Economics of Pace University's Lubin School of Business in New York City. Before joining the faculty at Pace University, Dr. Webster held positions as an international economist with the Central Intelligence Agency, the U.S. Department of Defense, Manufacturers Hanover Trust Company, and Continental Illinois National Bank and Trust Company. Dr. Webster has served as graduate and undergraduate finance program chair, and as faculty advisor to Beta Gamma Sigma, the international honor society for collegiate schools of business. He is the recipient of the Lubin School of Business Scholarly Research Award for Basic Scholarship, the Lubin School of Business Outstanding Faculty Service Award, the Pace University Award for Distinguished Service, and the Beta Gamma Sigma Commitment to Excellence Award. Dr. Webster received his BA from the School of International Service of American University, and his MA, MPhil, and PhD from the City University of New York.

Lightning Source UK Ltd.
Milton Keynes UK
UKOW02n1816240314

228719UK00005B/27/P

9 780739 186046